Bernard Beckett was awarded a Royal Society Fellowship for Teachers of Maths, Science and Technology in 2005, during which he worked at the Allan Wilson Centre for Molecular Evolution. During that year, he found himself looking for a way to combine his passions for science, philosophy and storytelling. *Falling for Science* is his first non-fiction book.

Bernard is a Wellington based writer. He's written several novels and won many awards for his fiction. His latest novel *Genesis* was also written as a result of his time at the Allan Wilson Centre. This 'otherworldly' sci-fi thriller was winner of the Young Adult Fiction category in the New Zealand Post 2007 Book Awards.

Also by Bernard Beckett

Fiction

Lester 1999

Red Cliff 2000

Jolt 2001

No Alarms 2002

Home Boys 2003

Malcolm and Juliet 2004

Deep Fried (with Clare Knighton) 2005

Genesis 2006

Falling for Science

Asking the Big Questions

BERNARD BECKETT

Longacre Press

I was lucky enough to spend 2005 on a NZ Science, Mathematics and Technology Teacher Fellowship. My thanks to the Royal Society, The Ministry of Research, Science and Technology and The Allan Wilson Centre, who respectively administered, funded and hosted this project. Thanks also to Mike Hendy from the Allan Wilson Centre, John Hannah of Victoria University and writer, Chris Else, of Wellington for reading my manuscript and offering advice.

ISBN 978 1 877361 72 2

A catalogue record for this book is available from the National Library of New Zealand.

First published by Longacre Press, 2007
30 Moray Place, Dunedin, New Zealand.

Book design by Christine Buess
Book cover design by Nick Wright
Printed by Astra Print, Wellington

www.longacre.co.nz

CONTENTS

'The fire burns the same way everywhere, but the law of the land differs from place to place.'

Protagoras, FIFTH CENTURY B.C.

'... when of a sudden I am surpriz'd to find, that instead of the usual copulations of proposition, *is* and *is not,* I meet no proposition that is not connected with an *ought* or an *ought not*. This change is imperceptible; but is however of the last consequence.'

David Hume, 1739

Introduction

It was tempting to give this book more of a best-seller title: *The Da Vinci Loud* perhaps, or *Men are from Earth, and so are Women: The Real Reason You and Your Mate Still Don't Get Along.* Certainly there are tales of ancient secrets in the pages that follow, not to mention those of battling sexes, great discoveries and enduring mysteries. But titles are not to be chosen lightly, for the way we name things often determines the way we subsequently think about them. In our modern world, when we distinguish between a fact and an opinion, or a theory and a speculation, when we label something a story, or call an explanation scientific, we are making a powerful set of assumptions. *Falling for Science* is my attempt to get hold of these assumptions and shake them about a bit.

I spent 2005 on a teaching fellowship at the Allan Wilson Centre for Molecular Ecology and Evolution. It is a place of excited industry, a hi-tech counterpart to the intrepid journeys of the Victorian naturalists. These days the strange new land being explored is the genome itself. DNA is extracted and sequenced; complex, computer-enabled mathematical models are constructed and critiqued; phylogenies are built, debated and dissembled. The workers in this field are explorers and they are story tellers, weaving together the fabric of an ambitious, compelling yarn. It was a natural place for the big question that drives this book to pop out. Just what is the difference between a story and a scientific explanation? And once this distinction is clarified, how do the two fit together? How can we use both, in our attempts to make sense of the world?

These are not abstract questions, for our answers underpin the way we respond to the religious, technological and ethical controversies of our times. It is my belief that we frequently get the answers wrong, thanks to a steady diet of oversimplification and wilful misrepresentation, and that's why I wrote this book. Plus I thought it might be fun.

I recently saw a documentary where a gentleman stared solemnly at the camera and pronounced, 'There is no external reality. The only reality is inside our heads.' Yet I wonder where the limits of this heroic conviction lie. Were I to take him to the top of a tall building and invite him to step off (because realities like gravity, kinetic energy and collision forces are after all, only inside his head), shouldn't I expect him to decline my offer? Sure, the picture of the world we carry inside our heads is just an interpretation of reality, but only in the banal sense that the meal we eat in a restaurant is the chef's interpretation of the raw ingredients. That doesn't make the food, nor indeed the external world, any less real. A chef can't turn a carrot into an apricot, and even a person in a deep hallucinatory state would have their experience altered by a high force collision with the tarmac. So to simply say, 'Science is storytelling,' and leave it at that seems to miss something important. For science, unlike other story forms, makes uncannily accurate predictions. Here we're talking about the sort of precision required to land a vehicle on Mars, zap a malignant tumour, or bounce a cellphone signal off a satellite. Things we haven't always been able to do, indeed once couldn't have even dreamed of doing. And that remarkable power needs explaining.

But, and here's where it really gets fascinating to me, the 'It's all just science, you know,' approach is just as problematic. The suggestion that we use story to fill in the gaps that science has not yet shone its light on is a natural extrapolation. Many things we once used non-predictive stories to explain (eclipses, plagues, floods) now have scientific explanations. Hence some choose to view science as a slowly encroaching tide, washing the shore clean of myth and superstition.

But while it might in principle be possible to reduce something as complex as a funny joke down to physics, I wonder if that might not rather ruin the moment. And as it is with laughter, so it is with love, justice, fear and longing. If someone tries to say to me that it simply comes down to science, it seems to deny the rather vital point that as a struggling Homo sapiens I desperately crave knowledge of things which are too complex, too subjectively slippery, too infected by humanity, to be captured by a mathematical formula.

Can science ever get inside this business of making meaning of our lives? Can it provide us with our ethical systems, or explain to us how to go about living a truly satisfying life? Don't we need stories for that stuff? I suspect we do, but suspicion does not an argument make. Hence the pages that follow.

It seems plainly silly to doubt the massive power of the rationalist project which day by day expands our understanding of the world, just as it seems a little loopy to deny the fact that we humans are in the end the weavers of stories, the creators of meaning. Yet both extremes are popular. We are exhorted to be either scientists or mystics, and upon claiming membership we are then expected to dedicate our energy to ridiculing the other side. I am sceptical by nature, and that makes me suspicious of clubs (indeed the sceptics' society always seems like a contradiction to me). Dull though many find it, it is the middle ground that most intrigues me. In trying to understand how science progresses, the strengths and accompanying limitations of its methodology, I am trying to understand the way science and storytelling ultimately must complement one another: to stake out a little territory for those of us who would like to see ourselves as well-informed fictionalists.

'What do scientists mean when they call something a fact?'
'What is thinking?'
'Do we have souls?'
'Is the current evolutionary model compatible with traditional religious beliefs?'

'Do we have real free will?'

'How much of the difference between male and female behaviour is biological?'

All of these questions are considered in this book. I'm not sure all of these questions have definitive answers; indeed by the end of this book you will see that I do not think some of them even make sense. But if they're the sort of questions which you find interesting, then maybe this is a good book for you. (And anyway, that's more than enough to read just standing there in the bookshop. The sales assistant is looking at you. Go on, put your hand in your pocket. If you don't like it, give it to someone you don't much care for at Christmas.)

Within these pages you will find tales of philosophers, from Protagoras and Plato in Ancient Greece through to those like Descartes, Hume, and Kant, who did their thinking in the seventeenth and eighteenth centuries. You will get a little of the methodology debates of the twentieth century, where Popper and Kuhn locked horns. And you will find many tales from science, particularly evolution. This is partly because this is the discipline I went swimming in during my research year, but also because evolution, more than any other discipline, highlights the way the model and the story can so easily become entangled.

I've set this book out in the manner of an argument, exploring the concepts in the order I had to get them straight in my head. First, what is knowledge, and how do we learn? Second, what can the history of western philosophical thought teach us on this matter? Next, how do scientists today see their work? With the theory behind me, I turn to some modern scientific controversies. I look at the history of evolutionary theory and the scraps that have accompanied it every step of the way. Then I move on to the contemporary debate surrounding attempts to use biology to explain human nature, and the fierce politics behind the old nature/nurture chestnut. For another example of storytelling and science working together, I consider the physical history of Pacific migration. And then I turn inward, looking at the way modern science informs and is informed by, the philosophy of the mind. What exactly is thinking? Can thought ever understand itself?

This is, to use the cliché, a book of two halves. I've called them 'Sailing into the Wind' and 'Heading Home'. The first part lays the philosophical groundwork, while the second is more interested in contemporary case studies. This is not an academic work. It is an idiosyncratic argument, one half of a conversation. My hope is that readers will agree with some bits and disagree with others, and by so engaging, alter their own view of the world a little. I have decided against footnotes, to avoid claiming for this work a status it does not aspire to. Let me however acknowledge the obvious. I could never have written this were it not for the tremendous talents of all the writers whose books and ideas I have plundered. To acknowledge this contribution I have included an appendix, where the books that had the greatest impact on my thinking are listed, along with a quick summary of what they contain, how they fit into my argument, and why you might want to consider reading them. This, I hope, is a useful sort of compromise.

To finish this introduction, let me show my philosophical colours from the outset. During the fifth century B.C., trade and colonisation opened up the Greek States to increasing contact with people from different cultures and traditions. It was within this context that the Sophists asked that most sophisticated of all questions: how, in the face of such diversity, can we know the way we see the world is the right way? Today, two and a half millennia on, a defining characteristic of our modern, internet-connected, information-soggy world is plurality. Never before have we been more likely to rub shoulders with people from different backgrounds, who don't see the world exactly the way we do. We can either take a moment to ask which views we are certain of, and which must be left open for negotiation — or we can bomb each other.

one

Part One:

Sailing into the Wind

Falling for Science

1 Meaning Machines

This book's central argument is a simple one. The things we know, and the things we think we know, depend upon a weaving together of science and story. That is, the stance we take towards our world depends upon both the 'facts' we have come to trust, and the stories we use to arrange and interpret them. To all but the fundamentally inclined, this will seem a little obvious, but the devil as always lies in the detail. I will suggest that when science fails to acknowledge its storytelling component it becomes not just hopelessly dry, but ultimately meaningless. And I will contend that when our stories deny the relationships which science has made reliable, they cut themselves off from their life source and wither on the vine. The trick then is to be able to distinguish between the predictive model, and the interpretive story; that we may avoid the twin pratfalls of the mystic who denies science's power, and the reductionist who can not see the wood for the trees.

Let me make two commitments from the outset, against which you can later judge me. First, I'm going to describe the practical implications of my argument as often as possible, so as to avoid that familiar, 'Yes, but what's your point?' feeling. Get that emperor out parading in front of the people, is the plan. Part Two of this book is devoted to showing how a good understanding of the way prediction and interpretation feed off one another can give us a new perspective on some modern scientific controversies.

Second, I'm going to get my definitions out into the open as quickly and cleanly as possible. It is always tempting to make one's case unassailable by hiding the key assumptions from view, in the hope that obscurity might become a placeholder for profundity. The cynic might suggest entire academic careers have been built upon just this device. In my argument I'm going to use words like 'thinking', 'knowing', 'modelling' and 'meaning' as if we can all agree exactly what they refer to. And that of course is a form of cheating, because each of these terms represents a portal into the world of philosophical disputes. It's difficult to acknowledge these disputes without having them hijack the entire case, and I'm certainly not going to provide the sort of rigorous definitions a professional philosopher might insist upon. Nevertheless I feel I ought to spend the opening chapter explaining to you roughly what I mean when I talk about thinking, modelling and interpreting. So, on with it.

I've spent most of the last fifteen years working in high schools, watching teenagers battle with education. It's an environment I know pretty well, that place where an optimistic curriculum collides with the seething, contradictory paths of adolescence. A place where idealism and cynicism make for strangely comfortable bedfellows. I figure schools might therefore be a good place to go searching for the answer to our first question: what is knowledge? What do we mean when we say we know something, and how do we go about learning new things? My approach, rough and ready though it may be, is to demand that any response I cobble together to those questions should make sense within the very down-to-earth context of the modern school. Otherwise it's just words.

If we envisage a school as a production-line — young student in, old student out — then most teachers would admit it is difficult to quantify any value we might be adding along the way. In some cultures this would hardly matter. We keep them safe, do our best, and most of them seem to turn out okay, so really what's the problem? We do not however live in such sanguine times. Educationalists and bureaucrats alike devise the most devilish systems of assessment, hoping to measure the work we do, and so define this business of learning. Some would claim that all this has achieved so far is a transfer of resources from teaching to reporting-on-teaching, and that the state we are now asked to aspire to is one of well-documented mediocrity. (This is not the place for that debate, I mention it only because I couldn't resist.)

Yet while it is difficult to measure the changes which occur in our students, it is harder still to deny them. The students who leave our schools have changed greatly from those we first met five years earlier. Most of them manage the difficult transition from child to part-time drunkard with remarkable aplomb. The luckiest of them head out into the world with confidence and enthusiasm. Despite all our efforts to choke education off at the photocopier, the little darlings are undoubtedly learning stuff. Imagine you are the parent of a thirteen-year-old about to enter high school. What do you hope they will know a little more about by the time they leave? Here are some of the things I'd put on my wish list:

- How to introduce myself to strangers, and establish myself in a new social grouping.
- How to make choices for myself, and stand by them in the face of pressure.
- How to compromise.
- How to read, write, and deal with number and measurement at the level needed in a modern society.
- How to take what I know and apply it to new things.
- How to stick at something that is hard.
- How to make myself known to someone I'm attracted to, and face the incumbent risks.
- How to bounce back from failure and disappointment.
- How to deal with the fact that I have more questions than the world has answers for.
- How, in the face of all the challenges above, to maintain an enthusiasm for living.

Yes, I know, I've missed out algebra, the Treaty, and static images. Never mind. The point I'm getting at is the skills and sets of ideas listed above are complex, and they have to be learned. Some of them lend themselves to exposition and practice in the classroom, but most do not. Students by and large are left to work it out for themselves, through a process of trial and error. Just like the rest of us. We spend our whole lives muddling through like this, refining our approaches and our theories on the hoof. For the human being, living and learning are so tightly entwined they are synonymous.

And this is exactly the process I want to examine. Not the teacher-mediated, face-the-front-and-remember-there's-a-test-in-two-weeks sort of learning, but the real stuff. The stuff we all do naturally, day to day. We're learning machines, and I want to build a simple model of how we get knowledge and what we use it for, something to have in the back pocket when I go tackling the bigger themes. Here I should point out that I'm making a distinction between knowledge and skills, and I'm doing it purely for the sake of focus. I understand there's a huge difference between *knowing* how to make a good impression and not tripping on your shoelaces on the way over; but in this book it's the knowledge part I'm interested in.

Let me first give you a definition for knowledge. It's not watertight, and like all my definitions I think of it just as a framework on which to build my argument. The philosopher Wittgenstein remarked that once you're on the roof you can kick the ladder away (although you have to really really like being on the roof) and I'm using the same dubious escape to keep things folksy.

We might think of knowledge as being composed of two parts. The first is a model of associations or relationships. What colour goes with lemon? What implications go with driving too quickly past a speed camera? What emotion goes with your family forgetting your birthday? This sort of knowledge can be thought of as a great network. Anything we know in this way can be broken down into relationships: of qualities, causes, categories, etc. That's how computers do so well at mimicking basic aspects of our behaviour, like playing chess, or throwing tantrums. Only we're not just machines. (Actually, some people think we are, but we don't need to argue with them, because who cares what a machine thinks?)

Why do I say we are not just machines? Because we do something *more* than just build up associations. We also develop attitudes towards them. We develop points of view. So perhaps I shouldn't say we are not machines, but that we are the sorts of machines that are capable of building meaning into our world. We don't just observe the rain and understand by way of association that rain will make us wet. We go further, imagining what the downpour must be doing to our hair, and how this is going to make for a very bad first impression … which in turn is such a shame, given that he sounded so lovely on the phone…

The hopes we harbour, the fears we swallow and the choices we make all tell a story of a world made up not just of facts, but also points of view. To be a successful human being (and here I mean simply one that is able to function) we need a brain capable of two quite different things. We have to gain some sort of understanding about the way the things we experience relate to one another (heavy things sink, hot things burn, wearing pink to the rugby gets me in trouble) and we also need a way of finding within these relationships a place for motivation, meaning and delight. There is a world of difference between understanding the physics involved in floating, and lying back in a sparkling bay, soaking in the sun and feeling at peace with the world. In very simple terms then, knowledge is the answering of two distinct but related questions. What physical relationships in the world can I rely upon? And what do I make of a world made up of these relationships, what does it mean to me? The first of these questions requires that we build up models of the world which allow us to make predictions. This, I will suggest, is the business of science. It specialises in prediction. The second question requires that we find a way of relating these models to our own sense of self, which I contend is the business of story. Story specialises in meaningful explanations.

This distinction is absolutely crucial for my argument, and while it sounds all well and good in the abstract, in the real world pulling the two aspects apart is a messy business. Dealing with just this messiness is the book's broader mission, but to give you a taste for the task, a couple of examples early on might help.

Take the familiar conflict between a father and his fifteen-year-old son. It goes something like this. Every morning father drops son off at school on his way to work. Son attends one of those liberal, non-uniform-wearing schools (you know, lax discipline and plenty of drugs) while father works as a mathematics professor at the local university. And every morning exactly the same conflict emerges. Dad is ready to leave and paces about, car keys in hand, while son, who had to be told to get up four times, is still trying to decide which shoes to put on, and even after that will not leave the house until he has attended to his hair. (How does his father not understand this? What is he, insane?)

In these tense moments, each is tempted to dismiss the other: as old and uptight on the one hand, as self-centered and indolent on the other. The question that interests me is how much of this conflict, which drives both of them absolutely nuts and is in danger of growing into a severe relationship meltdown, is based upon different sets of knowledge about the facts of the world, and how much of it stems from differing points of views? In other words, all the history and hormone aside, is this a dispute about physical relationships in the world, or the stories which surround them?

Undoubtedly it's partly about predictive information: what the two combatants imagine will be the implications of changing their behaviour. Son knows what it is like to arrive at school poorly dressed. He has seen the way others are judged, and knows how he does it himself. He knows how it feels to catch a glimpse of his own reflection on days when he's looking a little crap. And son knows things about style his father has never grasped. He prides himself on being able to judge what colours and cuts work together. He knows what is in, and what isn't. He keeps himself informed. Crucially, he also knows that the current object of his relentless affection will be sitting with him in English period one, a nugget of information he has absolutely no intention of sharing with the family.

And father too knows a thing or two about the world. He knows about traffic flows in the city, and the delays to expect this morning now that the gas pipe refittings are in full swing. He knows that staffing numbers are under review in his department, just as he is aware what a stickler for time his Head of Faculty is.

So it is quite likely that part of the conflict comes down to a lack of understanding of the other's world. But that's not all it comes down to. For even if father and son could sit together in the same room long enough to explain the details of their existence to one another, these new understandings would not of themselves shape the compromise. For son still thinks father could do with relaxing a little. After all, he does still manage to get to work on time almost every morning, albeit a little breathless. He doesn't have to stress, sitting in traffic. He could just turn the stereo up and chill out a little. In the son's mind, the real problem is his father's attitude. He's an order freak, and he needs to get over it. For his part, father has little taste for his son's appearance obsession. Just throw something on and get out there, father thinks. Hell, it's what he does: finds the first button-up shirt and sensibly elasticated trousers to come to hand. And in his day there was none of this sleeping in. Today's youth, father fears, have it too easy, and have been allowed to believe the world should be arranged purely for the purposes of their own comfort. What's more, father suspects a lot of the time his son just does it to wind him up. He's seen it in his eyes.

This side of the argument is not so much about the facts of the matter as it is about the stories behind them. It is about the motivations and qualities each ascribes to the other, and it is about social mores, the endless list of 'shoulds' that lubricate our human interactions. And as anyone who has ever tried arguing with a teenager will know, when such stories clash, there's no reasoning your way out of it.

This is what I'll be referring to, when I write about the difference between modelling and meaning. Our models tell us what to expect from the world, be it traffic jams or peer-derision, while meaning relates to the way we see our obligations, motivations and greater purpose. To be human is to make sense of the world by combining both aspects. (And yes, as we will see many times in this book, beneath it all lies our animal nature, the canvas of instincts upon which our brain-tales must be painted, but we'll come to that all in good time.)

What is true for the personal also holds for the political. Consider for instance the issue of global warming. Does it worry you? What are you doing about it? And isn't this in turn driven by two different levels of reasoning? On the one hand is the model you are trying to construct. Do you believe the planet is warming up? What do you think is causing it? Can anything be done? What measures are likely to be most effective?

But beneath this model sits your personal story. How much do you like driving your car? How much do you care about things that will happen after you are dead? How silly do you feel on a bicycle? How emotionally appealing are conspiracy theories to you? What do your friends think?

Public arguments on such issues often highlight the first while neglecting the second. Activists on both sides fall into the trap of believing everybody has the same values as them, and therefore it's just a case of sorting out the model. It is my solid contention that we humans just don't work that way. The desire to be reasonable is only one of our many motivations, and is not in the ascendancy as often as we might like to believe. Of course, the reason this is often skipped over is that few of us are brave enough to say out loud, 'I think you're right, I just don't much care.'

That's just a small taster of the sorts of things I have in mind when talking about the difference between prediction and interpretation. The chapter on the philosopher Karl Popper looks at the link between prediction and science far more formally. What we need next is a rough and ready description of the way we accumulate our knowledge. How do we get it, and why do we bother?

The obvious answer is it helps us survive. By forming reliable associations between events, qualities and the like, we can make predictions, and those predictions can help us out mightily. Popper said, 'We have ideas that they may die in our stead.' A lovely phrase. In practical terms, this means things like 'If I eat that berry, I will die', or 'If I say what I think of that dress, I will lose a friend.' A sophisticated model of the world allows us to anticipate and indeed manipulate events.

Such a skill is not of itself uniquely human. At the most basic level, much of our associative knowledge is instinctive. We breathe for instance, because our body in some sense 'knows' that not-breathing is bad. A fear of heights, being alerted by sudden loud noises, a desire to eat when we are hungry: these all reflect some sort of knowledge about the way the world is, and how to respond to it. In the same way, a rat has predictive knowledge that tells it, 'If I do not avoid cats, the outcome will not be favourable.'

Let me veer off the path for a moment to tell a cool little story. There is a clever little parasite by the name of *Toxoplasma gondii* that reproduces in a cat's stomach. It's well suited to this environment, but the digestion cycle being what it is, it is later expelled by the cat. The challenge it faces is simple: how to get from the cat poo back into a cat? Were cats filthy dog-like creatures, the solution would be obvious. The parasite however needs a middleman. Enter the rat, which feeds on cat faeces. And cats do in fact eat rats, so you can see how this might work, from the parasite's point of view. Any good system can be made great with a little extra design. To increase the odds, *Toxoplasma gondii*, when inside the rodent, alters the rat's brain chemistry, so causing it to be less fearful of cats. Seriously! It appears that the parasite is able to mess with that part of the brain which deals with risk-taking in rodents. They become less cautious, and hence easier prey for the favourite feline in your life. Beat that, pharmaceutical companies.

I tell this story not just for reasons scatological. As the 'we're just machines' lobby is fond of pointing out, at the instinctive level, we are in some sense programmed during development. There is something unmistakably mechanical about the whole process. And mechanical processes can be altered, just by adjusting an input. This is what happens in the case of our parasite. Without the parasite, the rat is a machine that avoids cats. With the parasite, it is reprogrammed. René Descartes was the first to propose such an automated view of animals: I'll have more to say on that later. For now, note that the same sort of reasoning sometimes works well with people. Without a testosterone surge, a little boy avoids girls. With the surge, and a few beers in him to boot, all bets are off.

Yet there is a crucial difference. Judges don't let criminals off on the basis that, 'You're an animal, and animals have trouble controlling themselves sometimes.' Nor should they. In a criminal court, the bar for establishing lack of control is high. (If you had control when you chose to take the drugs, then forget using that as a defence, too.) We choose to see ourselves as animals, and then some. We extend this description to our fellow human beings, offering them the highest compliment by crediting them with the ability to control or at least moderate their actions. Our educative, artistic, economic, political and legal systems are all based upon this premise. The odd cult may refrain from voting on the grounds that their god is in control of the outcome, but most of us only go to weirdville for visits. So let's stick with tradition on this one, and move beyond instinctive knowledge to consider other higher human uses.

Human beings, largely through the medium of language, have created a world of abstract ideas that makes our predictive models extremely subtle and sophisticated. We don't just know that fire burns, food sustains, or objects more often fall down than up. We also have theories to explain why combustion engines work, why Jenny is flirting with Sebastian, why other drivers are all sticking to the left hand side of the road, or why interest rates are rising. These too allow us to predict events and alter our behaviours. The models aren't always perfect, but we're used to this, and cope with change. When we're driving in a foreign country for instance, we can generally adapt to its different road rules.

We don't just use our models for practical ends though, and this will become important later when we consider science's contribution to our knowledge base. For reasons both cultural and biological, we have come to enjoy models for their own sake. People like putting things together, be it analysing clusters of stars, or team line-ups at The World Cup. We like gossiping, reading biographies, and collecting paraphernalia. Models that do a good job of relating together the things we observe are often very satisfying, even though the relationships will never be of any practical use to us at all. Schoolteachers are hugely grateful for this fact.

From here on in I am going to combine these two knowledge uses — modelling for predictive and recreational purposes — under the term modelling. I'm doing this purely for the sake of simplicity. Later, when we consider in more detail the definitions of science, I will pull these two modes apart again and consider the differences in more detail. But one thing at a time.

The other use of knowledge which we met with our feuding father and son — making meaning of our world — is not so straightforward. I'm going to turn to another philosopher to help me out: the existentialist Jean Paul Sartre, who once said:

> Philosophy is concerned with man — who is at once an agent and an actor, who produces and plays his drama while he lives the contradictions of his situation, until either his individuality is shattered or his conflicts are resolved.

As people, we are both blessed and cursed with big brains and big cultures. At some stage the universe found itself populated by creatures who could wonder what it was all about. Us. We are aware of our own mortality. We see ourselves as individuals. We grasp the need to know others, and in our darker moments wonder if this is ever really possible. We have complex language which allows us to communicate with one another, and with ourselves. We can think in abstract terms. As we go through life we tell ourselves and each other stories about all the things we see and do. We make decisions. And in order to make these decisions we devise systems of ethics, for ourselves and our communities. All this, as much as our genetic history and our surrounding physical environment, defines what it means to be a human being. I know I'm showing my novelist bias here, but to my mind any model of humanity that doesn't explain this is simply a poor model.

The next question is where do we get our knowledge from? I'd suggest four sources. Again, the categories are neither rigid nor exclusive; they're just a framework for thinking about the problem. We get new ideas directly from other sources: what I'd call received knowledge. Gossip from a friend, a formula copied down from the whiteboard, something you read in a book, saw on television, got off the net. We also get ideas through direct observation. For example, as children we do a lot of watching, picking up patterns, and building up categories and lines of causation as they learn to deal with the world. We get ideas through contemplation, by thinking about what is already in our head, rearranging it, and teasing out the implications. And finally, some of our ideas about the world we are simply born with: our instincts if you like. (At this point, I should declare my anti-mystical prejudices. I don't believe new ideas come to us fully formed from without. Ancient memories, voices from beyond the grave and chatty gods are all metaphors I struggle with.)

We are obviously in no position to take on board every new idea available to us. There are countless ways of rearranging the information we hold inside our heads. And every day we are bombarded with new information from without. Conversations we choose not to listen to, books we do not open, websites we do not visit, newspaper headlines we do not remember. Much of this is dictated by random circumstance, but at some level we are also making choices. Why do some new ideas fascinate us? Why are others too boring to even attend to, let alone remember? We are like window shoppers walking along a busy high street. We barely notice what is in some windows; others we note, but walk past; and there are those we linger over, taking in the detail of what is on offer. If we think again like a schoolteacher, students are remarkably discriminating when it comes to what they will attend to. The fidgety little boy will listen in rapt silence to the conversation of the two girls behind him, taking every care not to give the game away. Try to teach him a little Pythagorus and he's gone in sixty seconds, and the next thing you know, someone's diagnosed him ADHD.

So why the difference? I like to think of it this way. Not all knowledge costs us the same amount to accumulate, and not all knowledge is worth as much to us when we have it. 'Why do we have to learn this?' the teenagers cry, and reluctantly we have to admit, it's not a bad question.

Think about the cost of new information. How long will it take to get the concept? A quickly read billboard and a chuckle, or five long years of post-graduate study? And will it be worth the effort? The philosopher Daniel Dennett illustrates this point with a nice story. A person scales a mountain to speak to the guru. 'Tell me,' our climber asks, 'what is the meaning of life?' 'I can tell you,' the guru replies, 'but first I need you to go back down and spend long years studying logic and mathematics, that you might understand my answer.' 'Nah, doesn't matter,' the disappointed climber responds.

But it's not just time and effort. Sometimes, new things are easy to understand, but hard to accept. They take a lot of integrating. It's like getting the new couch home, only to find now you have to change the carpet and the curtains. How did people feel when they first discovered the earth was many billions of years old, and did not support literal biblical interpretations? Must have been tricky. Or that it was not at the centre of the universe, or that people share a recent common ancestor with chimpanzees? How do we feel when we discover the flaws in a personal hero, or a favourite incident in our past didn't happen exactly as we remember it? The sense of betrayal is a familiar one, and a harbinger of the hard work ahead, as we are forced to rethink a thing or two. There's that familiar feeling of resistance when we hear something we do not want to hear, something that challenges our prejudices. The instinct is to shut it off, shout it down, formulate rebuttals before we have even heard the full explanation. In politics this is called debating the issues. The lengths we will go to, to block out unwelcome information, is one of our defining, and less attractive, characteristics.

Not only does new knowledge come with different price tags, it comes with different levels of usefulness. Should the Government decide that from tomorrow we're all going to drive on the right hand side of the road, I would like to know about it now (and I'd want everybody else to know about it now too, to make it really useful). When, however, some lunatic on television screams to me that there is a one-off-never-to-be-repeated-vacuum-cleaner-liquidation-sale, I just wish he'd shut up.

Finally, in terms of both use and meaning, the ultimate shopper's trap lies waiting just around the corner. What if the couch that felt so good in the shop is dead uncomfortable after a couple of hours? What if the new CD you read a great review for turns out to be rubbish, or the mirrors in the clothing shop told big, fat, dirty lies? So it is with knowledge. The worth of a new idea can only be judged in time. You hear something interesting, which modifies a point of view you hold. You enjoy this new insight. Later, you share this with a friend, who points out a flaw in the thinking. Or you notice the flaw yourself first (phew), or you don't notice it exactly, but somehow the idea just loses its appeal for you. You forget about it. Predictive value is even harder to anticipate. Upon arriving in a new country you have the road rules explained to you. You naively predict this is how people will behave on the roads. After an hour of driving, you are well and truly disabused of this notion, and you too sound your horn loudly before proceeding at pace through traffic lights, green or otherwise.

Uncertainty means short cuts. Rules not so much for knowing, as for guessing. We learn to trust some sources more than others. We develop prejudices, frameworks upon which to hang new knowledge. We cultivate some interests, and ignore others. We may take the expedient decision to follow the crowd, figuring that most people will be mostly right most of the time. (Everybody is running your way shouting fire! Do you run with them, or go and check for yourself?)

We may adopt a conservative stance, placing a higher burden of proof on new ideas than on existing ones. And most of all we set limits on our own curiosity. At some stage, we stop asking why, and simply get on with the business of living.

And that, in a hand waving sort of a way, is what I mean when I talk about knowledge in this book. I'm talking about the things we observe and the way we put those observations into models, so giving us the ability to make predictions and manipulate our world. And I'm talking about the way we interpret these models, to fit in with the larger narrative of our own lives. In short I'm talking about the way we remarkable machines construct meaning. In this book, the distinction between prediction and interpretation is going to be vital, which is why I wanted to get it out in the open from the outset.

That's all I want to say about definitions for now. We are ready to move on and look a little more closely at where the special type of knowledge we refer to as science comes from. How exactly do scientists know the things they know? Can science ever show us the world as it actually is? Is it even correct to speak of scientific knowledge as being different from other forms of knowing? These are the questions I want to address in the next two chapters, and that means we now have to delve back into history, to see how some truly remarkable thinkers have approached exactly the same questions.

2 From Faith to Sophistry and Back Again

In the last chapter I suggested that, instinct aside, there are three sources of ideas: authority, reason and experience. To understand the history of scientific endeavour, we need to look at these categories a little more closely.

There are two distinct attitudes we can take towards ideas which come to us by way of authority. Think about how you might have reacted to the book so far. On the one hand, your attitude may be one of complete credulity. (I hope not.) If it is though, you will have early on decided, 'This person knows what they're about. So this is all true. Excellent.' This is the blind faith attitude, remarkably efficient, although obviously a tad risky.

Or you might take a more questioning approach. Maybe you still buy the argument, but only because there are signs that I've done some of the hard work for you; it looks like I've been through the reasoning, or at least chosen reliable sources. In this case, when you come across something that seems a bit wrong, you're likely to challenge it; take the reasoning apart, trace the sources, construct the counter argument or look for a second opinion. From this viewpoint, all information is contestable, and all argument ultimately can be broken down to observation and to reason. Authority becomes a shortcut to reason, not a replacement for it.

A nice way to think about this is in the changing relationship between a child and the adult world. A young child accepts information at face value. Adults are treated as authorities: people to turn to when something puzzling appears. A young child will believe the most delightfully outrageous stories if they are told by an adult; the one about an old guy coming down the chimney at Christmas is an obvious, non-controversial example. But at a certain age children begin to temper their credulity. They begin to seek out inconsistencies in the stories of their elders: contradictions in the reasoning, or thorny exceptions to the so-called rule. They turn from receivers of stories to challengers of them. In short, they become teenagers.

The ancient Greeks then, were responsible not only for democracy's birth, but also philosophy's puberty. Thales, who lived in the seventh and sixth centuries B.C., is often credited with kicking off the philosophical revolution. He devised a theory that the world was ultimately made of water, which on the face of it may appear a less than promising start to the tradition of rational enquiry. To understand the profundity of what he was doing we have to look a little more closely at his methods.

What was it that encouraged him to ask the question 'what is the world made of?' in the first place? Can't dirt be made of dirt, grass of grass, and goats of goat stuff? Why try to break it down further? Well, observation of the world shows that such substances are not fixed. Things rot, they burn, they corrode and they dissolve. This is the first hint that perhaps the stuff of which the world is made is not the stuff which is immediately apparent to us. The curiosity required to follow up on this hint gifted us the traditions of chemistry and physics, and for this alone Thales deserves considerable credit. It is precisely this curiosity, the intellectual restlessness that can not resist tugging at the loose thread of inconsistency, which drove the discoveries of ancient Greece.

But Thales was more than just curious, for consider the next obvious question. Why water? Why not fire, or flesh, or earth? Thales probably observed that things often dissolve in water, that water itself can turn solid when frozen, or evaporate when heated. Perhaps he also noted that things often liquify when rotting. Whatever his path of reasoning it seems likely he used observations such as these to edge towards his hypothesis. The leap to the declaration that all is made of water feels a little flimsy by modern standards, but again the first seed of a scientific method is visible. The need to make a theory consistent with our observations is absolutely crucial to model building. Every time we shrug and say, 'Who knows, it's just a mystery,' or 'Well it's obvious isn't it, at this point a miracle occurs,' we are constructing roadblocks to our knowledge. Thales encouraged us to dismantle them.

To curiosity and observation we must add a third feature: contestability. This is perhaps the most important of all. Consider the response an evidence-based hypothesis such as Thales' water theory invites, nay demands. By standing up and saying, 'Here is what I believe to be true, here is my evidence and here is my reasoning,' Thales was challenging others to check both the data and the logic behind the proposal. He was inviting criticism.

And criticise they did. It wasn't long before flaws were being pointed out. (If everything is water, how come some things float, and other things sink?) And human beings being what we are, the investigation doesn't end there. 'Okay then,' the critic is challenged, 'you're so clever, what do you think the world is made of?' A cycle of model, criticism, refinement and counter model is started; and two and a half thousand years later it shows no sign of slowing down.

Contrast this with a society where wisdom and authority go hand in hand, where things are accepted as truths because the anointed ones tell us so, and the importance of Thales becomes clear. Suddenly people are no longer being treated as children, asked to simply listen, accept and learn. The competition is no longer about the holiness of the source, but the quality of the justification. So our justifications improve, and our discoveries gather pace. In the language of the time, the world had moved from mythos to logos. And with the arrival of logos a new type of curiosity based upon observation, reasoning and intellectual competition was born.

That at least is the slightly grand telling favoured by introductory philosophy texts trying to lure students away from the temptations of computer animation. It doesn't take too much thought to realise hypothesising and experimentation were an essential part of human existence well before Thales came along. (Perhaps if I make her laugh … hey, she just touched my arm, excellent!) Nevertheless something important changed with Thales, as can be seen by the unprecedented intellectual fecundity of those who followed.

I have neither the space nor the expertise to do justice to the dazzling achievements of the great thinkers of ancient Greece, so allow me to highlight just a few, to give you a sense of the wonder of those times.

Heraclitus, building off the observation that the nature of things changes over time, introduced the idea that all is in a state of flux, that change is the one true constant. That is, the nature of the world as we observe it, with its fixed qualities and relationships, is something of an illusion. Over time, cliffs fall into the ocean, buildings crumble, people grow old. Again it is not difficult to see the train of thought which is set in place by such observations. What are the forces behind this change? How is it that some systems appear to be in balance? How can change be measured and predicted? From our privileged vantage point, the anticipation of modern physics is striking.

Democritus too seems to have been remarkably prescient. He suggested that all matter is made up of tiny, unseen particles, which combine together to give the things we see their familiar forms. He called them atoms. I want you to stop and think about that for a moment (not the name part, that's not quite as spooky as it seems, right?) but having the idea in the first place. What sort of inspired guesswork are we talking about, to take us from the world we see, to the world of atoms? Isn't that just stunning?

I particularly like the story about one of the early problems the atomic theory ran into. If everything, including ourselves, is just made of tiny particles, then how are qualities like colour, taste or sense transferred, people asked? How do I know that wall is red, not white? A fair question, and in response came the suggestion that other tiny 'quality atoms', signifying colour, sheen, smell, sound, whatever, travel to our eyes, nose or ears, where they are interpreted. But not so fast mister atom, came the reply. Once a redness atom gets inside me, and excites the 'I see red' atoms inside my brain, how do I know what's going on? How do I know to see red? As far as I know Democritus failed to provide a compelling answer to that question. Then again, so has everybody else. Perception (as opposed to the physical processes associated with perception) remains something of a mystery, and two thousand years later led Descartes to his (in)famous dualism. These days some theorists think they have an answer, while others believe the question to be unanswerable. We'll come back to that in a later chapter.

Another great example of the ingenuity of the ancients is Eratosthenes, a Greek mathematician who lived in Egypt in the third century B.C. Eratosthenes still earns mention in mathematics classes for his method of finding prime numbers. He was also a great one for a practical experiment. He heard of a particular well in what is now Aswan, where, at noon on the summer solstice, the sun shone directly down to its bottom — which is to say, at that point, the sun was exactly overhead. Now you and I, having heard such a thing, might think, 'Oh right, interesting' and turn back to other things. Eratosthenes, by contrast, thought, 'I can use this to show the earth is round, and what's more, I should be able to work out its circumference too.' It always makes me feel a little intellectually lazy when I hear of such people. Here's how he proceeded.

Wells are built to go straight down, which sounds a tad obvious I know, but consider the fact that if the angle is just a little bit off, it's not heading straight for the earth's gravitational centre, and when you lower the bucket it's going to bump against the side, and you'll end up with a lot of grime in your water when you pull it back up. One way of getting this right when digging wells is to hang a plumb bob (a weighted string) down the middle to guide the way. With this in mind, we can now think of wells as tunnels heading from the earth's surface straight towards its centre. Think of it as a pin stuck into an orange, aimed at the very middle.

Now, because an orange is round, two pins just a centimetre or so apart will not be parallel if they are both aimed towards the centre. Eratosthenes realised the same would be true for two wells. He went not to the well in Aswan, but to another in Alexandria 600 miles to the north. At noon on the summer solstice, he measured the angle between the light from the sun and the well, realising that if the earth was round like an orange, the sun would not be shining straight down this well as it was at its Aswan counterpart. He used this angle, and some basic maths (draw yourself a diagram, you'll be able to work it out), to calculate the earth's circumference as 45 000 kilometres. It is actually nearer 40 000, but for someone to get pretty much within ten per cent without leaving Egypt is I think an impressive feat.

This story gives us some good stuff to think about, so I'll stay with it a little longer. First, where did the hypothesis come from? Why did he think the world was round in the first place? It was a common supposition at the time. The early Greeks had noticed that when ships appeared on the horizon, they always saw the masts first, and thought a curving earth might explain it. So the roundness of the earth becomes a great example of what I mean when I talk about knowledge modelling the things we observe. And the thing about a model is that it doesn't just give an explanation (ah, so that's why we don't see the bottom of the ships) it also allows us to make powerful predictions (sail away long enough, and you'll be on your way back home). Crucially, some of these predictions may lend themselves to simple experiments, in this case the prediction that two deep wells sufficiently far apart will not be parallel. If this prediction turns out to be true, we begin to feel more confident of the model, even though later the model may be refined or replaced (the earth is not exactly round; it is deformed by its spinning!).

There is a lot more to say about the scientific method, and I'll get to it in time, but it is just good to note that the basic technique was in place well before anyone started calling themselves a scientist.

Of course, wonderful though all this argument and counter argument, speculation and refutation was, the culture of open enquiry comes with a downside. The only way we can make room for discovery in our lives is by clearing out a whole lot of certainty. And certainty is a deeply attractive quality, as anyone who has ever tuned in to talkback radio can attest. 'Sort of', 'in a way', 'perhaps' and 'as far as we know' are not our sexiest phrases. In an age where new and speculative models were being proposed and overturned at a tremendous rate, it was natural to wonder whether all this jostling was heading in a useful direction. Why replace mythology with enquiry if all our enquiries deliver up is uncertainty? If today's theories are destined to become tomorrow's curiosities, what then of the truth? Are we destined to never know anything for certain, and is that too high a price to pay?

In the words of Xeophanes (530 B.C.)

> And if by chance man were to utter
> The final truth, he would not himself know it;
> For all is but a woven web of guesses.

One group of early Greek thinkers, the Sophists, confronted this dilemma head on. By the fifth century B.C. the Greek city states were involved in a process of colonisation, and people were increasingly being exposed to foreign cultures, with their many different ways of thinking about the world. It is a measure of the progress Greek thinking had made in the one hundred and fifty years since Thales that the Sophists asked, 'What if they're right and we're wrong?' From here, it was only a short leap to 'What if none of us are right? What if truth, like beauty, is in the eye of the beholder?' Or to paraphrase the great sceptic Protagoras, 'What if man is the measure of all things?'

The Greeks had arrived at the doorstep of a very provocative idea: reason might lead us not to truth, but to interpretation. This sceptical viewpoint, that ultimately we can know nothing, that our truths are constructed by us, for our own convenience, deserves to be seen as the third pillar of the philosophical temple, holding its place alongside mysticism and rational enquiry. It is less comforting than the others — our certainties provide us with a place to stand in the world — but on purely logical grounds it has always presented a formidable challenge. How is it we know the things we do? Why do we believe our truth is the only truth?

The mystics will answer they are confident of their truth because it has been revealed to them, and the rationalists will back their truth because it is the place their reason led them. But the question then becomes obvious. What to do when two people's revelations contradict one another, in the case of the mystic? And in the case of reason, where are we to find the irrefutable premises upon which our reasoning is to be based? These are the key questions the sceptics introduced to the debate, and as we will keep seeing during this exploration, they're important.

All this brings me quite nicely to Pythagorus; yes, the triangle guy you learned about in school. Pythagoreans (he created his very own sect) were a curious mix of the rational and the mystical, and in some ways set a bad example from which we've never really recovered. The Pythagoreans believed mathematics could lead them directly to the sort of certainty people craved, a certainty built not upon revelation, but on reasoning. And to this day you will find physicists who go all misty-eyed at the ability of mathematics to shine light on hitherto unknown aspects of the physical world. Science has always attracted its fair share of mystics.

For the Pythagoreans, number was the thing. By number, they meant countable quantities (like five or 32) and ratios formed from them. They observed that many natural phenomena, for example musical harmonies, proportions in the body, and architecturally pleasing forms, could be described very accurately by numerical ratios. From here they leapt to the slightly spooky conclusion that perhaps number could be the key to unlocking nature's secrets. If only they could nut out the numerical patterns all things were based upon, they figured, the deeper truth of the universe would be revealed. Notice two important urges being combined here, those of reductionism and mysticism. Get anything down to its simplest model, and you can understand what it's really all about.

One of the great gifts the Pythagoreans left us was that of mathematical proof. Unlike the other philosophers, who posited theories which in time could be replaced by better ones, the Pythagoreans demonstrated that some mathematical truths are indeed eternal. Once proven, a mathematical statement can never be unproven (although an error can be made in a proof, which may later be spotted). The great irony here is that their new weapon was eventually turned on themselves. They managed to prove that the square root of two (which they knew existed conceptually as the length of the longer side of a unitary isosceles right angle triangle — sorry) was in fact irrational. That is, it could not be expressed as the ratio of any two natural numbers. To us this is hardly the most devastating news (fourteen-year-olds barely bat an eyelid.) But to the Pythagoreans, believing as they did that all things could ultimately be described using numbers, the irrefutable demonstration that there existed a theoretical length which could not be expressed as a ratio of two numbers was crushing. There is a story, probably apocraphyl, that a follower was executed for discovering it.

The spirit of the Pythagoreans did not however die with this discovery, thanks to the mighty contribution of Plato, perhaps the most famous of all philosophers. We'll get to him in a moment, but first let's just take a moment to draw breath, lest all this ancient talk is beginning to move things out of focus.

I am interested, here, in the traditions of thinking that led to our modern picture of the world. In particular I'm fascinated by the balance between knowledge through reason and observation (what would in time be called science) and knowledge through imagination or inspiration. By the end of this book I will have claimed that the two forms of knowledge complement one another, providing both our models (science) and our meaning (story). This however requires a separation between appearance and interpretation, which in itself requires an acceptance of at least some of scepticism's arguments. For now though I just want to show how the birth of the scientific thinking led to the very argument I am trying to get to the bottom of; the way our observations, our beliefs and our doubts dance together in the quest for meaning.

With the ancient Greeks, we saw the birth of the scientific method, characterised by curiosity, argument, change and, as in the case of Eratosthenes, some remarkable practical achievements. Nevertheless, from the outset it was always a moot point whether these new models could provide for us the sort of meaning we crave. Could these new insights into the physical world help us cope with our deepest hopes, fears and insecurities? Could they help us cut something magnificent from our human cloth?

And that's the challenge Thales-type thinking introduces to the world view. In a world where nothing new is being discovered, it's easy for our stories to appear solid and timeless. Science however by its very nature challenges old assumptions: it is a machine that feeds off change and uncertainty. But what do we do in the face of a machine which undermines our stories and replaces them only with models of measurement and prediction? Where do we turn for our greater meaning? The ancient Greeks were the first people forced to face this daunting challenge.

One approach is to follow the Sophists, and argue that all is but a woven web of guesses, that notions of truth must be discarded. Science can produce models which predict the world, but it is still up to us to breathe meaning into these relationships. And the way we choose to do that is in some sense arbitrary. Truth is not our guide so much as our comforter.

But the problem for the Sophists is ultimately one of marketing. For many people, sophistry isn't all that satisfying. If we really are just choosing stories for their utility, then we can legitimately ask how useful a story sophistry itself is, for any philosophy must live by its own rules. The young debater quickly learns a plausible argument can be constructed in favour of almost any case. For a while this makes them feel awfully clever. But in time this triumph of shallow artistry begins to fade. While many of us may enjoy constructing argument for argument's sake, there seems to be a shared urge to dig below the sophistry and discover, well, the truth; that great, unchanging, informing perspective which will make our lives somehow meaningful. Once we believe our stance is arbitrary it's in danger of losing its appeal. And that brings me nicely back to Plato, who knew a thing or two about marketing.

Plato offered respite from all this accursed doubting. A supreme communicator with a poet's instincts and a knack for striking, timeless imagery, he had no time for ultimate uncertainties. Born into an age of political turmoil, Plato perhaps reflected the mood of his generation, a great desire to find something solid to cling to. He was influenced by the Pythagoreans, and understood full well the implications of the discovery of irrational numbers. So the truth isn't number then, he concluded. Oh well, no point giving up now. Perhaps it's shapes! Against the sceptical point of view, he sought to re-establish the quest for the deeper order of the universe. His lasting influence was the body blow he dealt the sceptics, one it would take them two millennia to recover from.

Plato replaced the Pythagorean concept of eternal numerical relationships with that of eternal forms. The truth of the world existed in the shape of these perfect forms or ideals, he suggested, while the world as we observed it was no more than a poor shadow of this greater reality. He used the famous image of people in a cave, forced to view the world only through the shadows it cast upon the cave's walls, while outside the truth was available only to those philosophers who dared step into the light. Immediately the shift in methodology is apparent. Plato didn't base his theories on close observation of the world, for such a world is imperfect and unreliable. The truth, as he saw it, was beyond the senses, and could be known only through deeper contemplation.

Plato didn't abandon reason altogether, indeed he accepted that rationality could take us some of the way and analysed this path of reason in much detail. But he distinguished between reason and insight, which came from some more mystical source. The business of being human — according to Plato, our very duty — was to seek truth, beauty and justice, which were not defined by humans, but existed as eternal, unchanging forms. Ultimate truths, if you like. Life became a quest for meaning. Truth wasn't just some story we told ourselves as the Sophists would have it, for it existed independently of us, constant and eternal.

And if the Sophists suffered a product that was market-unfriendly, Plato had no such problems. For not only was he offering stability and certainty, but his followers got a little eternity too, as part of the bargain. For Plato posited the existence of the human soul: that ultimate comforter, freeing us from the constraints of death. His reasoning went something like this. All earthly things are imperfect, so this must include our bodies. Therefore our true nature must be non-physical, for our true nature is perfect. The other option, that we are simply imperfect and incapable of knowing anything but our own cobbled together versions of the world, was pushed aside. It's not sexy enough, as stories go. It doesn't make us glorious.

So in the ongoing jostling between mysticism, rationalism and scepticism, Plato's legacy was to promote the first and banish the last, while sending out contradictory messages about the extent of rationalism's power. Aristotle, the other great inspiration for the Christian tradition, subsequently added twists of his own. One aspect of the Aristotelean tradition is worth highlighting briefly, simply because it is an idea which we will see coming back to the surface in later evolutionary debates.

Following Plato's lead, Aristotle encouraged us to think of the material world in terms of its causes. The best way to understand the nature of anything, he suggested, is to ask what it is for.

When we examine something, Aristotle thought, we should ask, what is the purpose its creator had in mind? Just as we can't properly understand a vase if we do not know it was designed to hold water, neither can we truly understand humanity without some notion of its greater purpose.

The trouble with this stance, as with Plato's, is that it requires a method by which the higher truths or powers can be discovered. Plato thought contemplation was a big part of this, and both he and Aristotle were confident that understanding the movement of the stars was key to it all, but there was still something missing before it could really take off. One can imagine a modern day advertising executive surveying the potential of the brand and saying, 'Well, you're offering comfort and certainty, which is great in the abstract, but people want something a little more concrete. Something they can read perhaps, a guide to how they should live. Some simple answers.'

By the time of Christ's birth a good deal of the Christian tradition was already in place, thanks to Plato and Aristotle. We had the notion of eternal truth, of a prime mover, of knowledge by way of revelation, an eternal soul and a perfect heavenly realm. This set of beliefs, combined with Christ's teachings and the Judaic tradition, went on to provide the West's intellectual framework for the next fifteen hundred years. Story was again in the ascendancy. If you wanted to know what it meant to be human, how the world worked, and how we should act, it was simply a case of consulting the man himself by way of the scriptures, with a fair amount of guidance from those trained in the art of interpretation. Thales' urge to look more closely at the physical world, to seek explanations for the puzzling and the inconsistent, had flowered only briefly. Why try to nut things out for ourselves when God has given us the answers directly? And the great sceptical challenge was equally quashed by this divine intervention. The truth was out there all right, it was all just a matter of opening our hearts to it.

I said earlier that certainty in our stories is often undermined by new discoveries, which force us to rethink our central metaphors. In this respect Christianity had something of a dream run, in that the Dark Ages aren't exactly known for their long list of intellectual advances. But Dark Ages don't last forever, so let's leap from ancient Greece to the birth of the modern era, where the tale of story, science and scepticism gets interesting again.

There are any number of candidates to stand as the Thales equivalent of the modern age. Perhaps it should be Copernicus, for moving the earth away from the centre of the universe; perhaps the fall of Constantinople and the revival of interest in the ancient texts and methodologies; perhaps the flow of mathematics and technologies in from the east. I'm going to choose Galileo (1564–1642) and Francis Bacon (1561–1626) as my rather arbitrary figureheads, because a story has to start somewhere and they represent so many of the things I want to talk about. They were contemporaries of Shakespeare's, which gives a nice cultural reference point (Galileo was reputedly born on the day Michaelangelo died, for those who like such trivia). So, how exactly did the scientific revolution get its start?

Shakespeare's time was, by modern standards, a time of superstition. Life for most was 'nasty, brutish and short', to misplace Hobbe's immortal phrase. Authority was received from on high, via the anointed guardians of wisdom, the educated interpreters of the scared texts. Supernatural forces were taken for granted, and their mysterious ways were not open to human scrutiny. Over the next two hundred years, rationalism would rise up and transform the world, and the sceptics would once again have cause to ask their pesky questions. Many excellent histories of this period have been written, and this isn't one of them. All I want to do is pull out a few key figures in order to show how science, story and scepticism once again struggled for their place in the sun.

Galileo's story is well known. A fine mathematician, he made use of advances in telescope technology to confirm Copernicus' belief that the solar system was centred about the sun, with the earth circling it as it spun daily on its axis. As is now legendary, factions within the Catholic Church made a seriously bad call in opposing this theory as heretical. It would be easy to say this was ignorant superstition opposing serious scholarship, but that is perhaps to be too wise in hindsight. There were a number of factors behind the dispute: personal, political, scientific and theological.

On the personal front Galileo was by all accounts a prickly soul who had the knack of upsetting important people like the Pope. Politically speaking, the Catholic Church at the time was going through a particularly conservative phase, in a reaction to the rise of Protestantism. Scientifically, Galileo's work was new and the jury was still out. It's easy now to see he was right, but sorting the visionary from the crackpot is harder than you might imagine. And theologically, accepting Galileo's take on things was going to involve some serious reformulating of the dominant church mythology. An institution that rules through the myth of divine guidance doesn't take kindly to revisions forced upon it by mere observation. Still, the risks involved in not accepting Galileo's version were also large. If he was right, then the church would face the double challenge of explaining the gaff, and the subsequent persecution of the truth-teller. He was, and history has judged the church badly because of it.

What this incident demonstrates most clearly is a changing of the guard. Galileo's challenge was only one of many; a torrent of new models was about to tumble down upon the world. If theologians wanted their institutions to survive, they were going to have to adapt more flexible storytelling strategies. It was certainly going to be harder and harder for church authorities to present themselves as the last word on all things. But what was fuelling this new age of discovery?

Attributing causes is more an art than a science, but in the interest of storytelling we can say that in part it came down to the very methodology Galileo employed. The telescope here becomes a useful metaphor, for it is an instrument designed specifically for careful observation. As with Eratosthenes at the well, Galileo turned to precise measurement as the launching pad for new theories. Notice that this runs counter to the Platonic intuition, that measurements of the imperfect natural world could not lead to any understanding of deeper, more perfect realities. The new generation of theorists were distinguished by their willingness to dirty their hands on real world data, and their belief that nature will offer up its secrets to those who are careful enough in their observations. Galileo was a pioneer of this approach, dropping things from towers, tracing the trajectories of cannon balls, timing pendulums and making careful measurements of the stars. If you want to know how the world works, was the brave new motto, then you ought to start by watching it very, very carefully.

If Galileo was the exemplary practitioner, Francis Bacon was the chief cheerleader. He, perhaps more than any other thinker, captured the optimism of a new mood at the opening of the seventeenth century, sure as he was that all of nature was bound by simple laws. Bacon was convinced that the careful observer and trained reasoner could uncover these eternal relationships, and he urged people

> to make peace between themselves, and turning with united force against the Nature of things, to storm and occupy her castles and strongholds, and extend the bounds of human empire.

By prising nature's secrets free, Bacon thought, we could tame her. Bacon is referred to as the father of induction, and here is a good place to linger a little longer on what that is all about.

Induction is the principle by which local observations are extended to form universal rules. There is nothing particularly strange in this, it's something we all do naturally. Having observed that items we throw into the air eventually come back down, we are happy to generalise and expect this to be some sort of law-like property. Over time, only the particularly dim witted or philosophically obstinate remain surprised by the continuing operation of this rule. Equally, when negative evidence comes in, we begin to suspect that the law is not all it is cracked up to be. This too is natural. (You told me a pink sunset means good weather the next day, but there was one last night, and now it's raining. Your rule is bollocks.)

Bacon attempted to formalise this way of thinking. He wanted to move past the obvious usefulness of inductive thinking, and ponder also its meaning. What is it about the nature of the world that delivers up such regularity, he wondered. His hunch was that Nature itself consisted of a series of simple rules, and that by understanding these rules we could come to understand the very nature of our existence.

It is perhaps not too much of a stretch to suggest that in doing this he established the modern scientific agenda.

Bacon also had a lot to say on the methodology that could be used to uncover these rules. He preached careful observation and experimentation, and suggested that the proof of a hypothesis could be found in the predictions it made. This insistence on tying philosophical speculation to physically reproducible results marks Bacon out as a tremendously important thinker in the history of scientific tradition. He considered the difficulties of objectivity at length, and wrote on the many traps the careless mind may fall into.

Bacon also attempted to establish a line of demarcation between rational philosophy and theology. Apparently a man of orthodox faith, he considered that the careful method of enquiry he outlined had no power to uncover the deeper mystical truths of the universe, the nature of God for instance, or the purpose of human existence. And by implication, he did not believe articles of faith could be much help in revealing the nature of the physical world. So, to function fully as a human being, two ways of gaining knowledge were required: a scientific method, for understanding the physical world, and some faith aspect, to reveal its meaning. Of course, this form of double thinking was easier to sustain four hundred years ago. Bacon could never have imagined just how far science would go in claiming the ground upon which religion stood. Evolutionary theory does away with the need to explain away nature's complex design and modern neuroscience is steadily chipping away at the mysteries of thought. So it is often tempting for the modern scientist to believe that there is no place for interpretation, and that in the end all can be reduced to predictive models.

It is not difficult today to find those who quite casually move from scientific model to cultural interpretation without so much as blinking, an attitude I wish to take to task in the second half of this book. As a small appetiser, consider one of the most famous communicators of modern science, biologist Richard Dawkins. In an impassioned essay on the rightful place of the great apes, Richard Dawkins reminded us we could assemble a line of creatures running all the way from chimpanzee to human (by way of our common ancestor intermediary) where each creature in the line was practically indistinguishable from the one next to it. While this is undoubtedly correct, his conclusion that this insight should inform the way we see the other great apes is dubious. After all, the same is true of the human and the carrot. It would be premature to dismiss Bacon's distinction between the realms of science and imagination just yet.

As we move through the scientific century, there are two more names that simply must be mentioned: Renée Descartes (1596–1650) and Isaac Newton (1643–1727). Like Galileo and Bacon, their impact on the way we see the world has been long lasting. It is telling of Descartes' influence that it is difficult to know where to start with him, and equally telling that his legacy is a series of puzzles which we are yet to dispose of. Perhaps Descartes' greatest contribution was the stance he took regarding the natural world. He believed, like Bacon, that all of nature was simply the unfolding of a giant clockwork machine, working according to the rules of simple, underlying forces. He posited that even animals were nothing but automata, and that what seems to us to be a mysterious life force could be reduced to the same elemental conditions controlling the turning of the tides or the spinning of the night sky. In this sense he was the arch determinist, and so defined the true scope of science's ambitions. If only we could understand the rules, the mechanics of existence, we could understand everything. Well, almost everything, for Descartes saw fit to call for one exception.

To understand the exception, we need better to understand Descartes' methodology. Unlike Bacon, he was no empiricist. He trusted abstract reason more than careful observation, a stance perhaps aided by his great skill as a mathematician. One of the things Descartes dreamed of was a system of reasoning from first principles which could in turn yield all human knowledge. But what to use as a first principle? Descartes attempted to adopt a position of extreme scepticism. What is it I can trust, he asked? What is it about the world that everybody can agree on, and so can be used as the launching pad for a rationalist philosophy? He concluded data collected by his senses could not be trusted, for any evidence might well be planted by a malevolent spirit, intent upon tricking him. In time he was even able to doubt the existence of his own body, for what if this too was just a figment of his imagination? (Perhaps I don't have an arm at all. Perhaps I just think I do!)

This sort of doubting inspired the later idealists, who liked nothing better than to argue nothing at all really existed. How can we really know it isn't all just an illusion? Think the seventeenth century version of *The Matrix*. Most sane people will sympathise with Samuel Johnson's response. 'I refute it thus,' he is said to have uttered, before kicking a rock. ('Ah yes,' replies the nutter, 'but perhaps you just imagined you kicked the rock.' 'But you saw it.' 'Perhaps you've imagined me too.') Or Jonathan Swift, who heard his idealist friend knocking at the door, and declined to open it on the grounds that 'The door is in your head. You unlock it.' This again underlines for us the dubious worth of theories which are in principle feasible, but in reality rather useless.

Descartes though was not an idealist. He thought he had found something which could not be doubted. He concluded the one thing he could be sure of was that he was a thing that thinks. He reasoned that while he could say of his body, 'Perhaps I just think I have a body', the equivalent 'Perhaps I only think I am thinking' makes less sense. 'I think therefore I am' is the lasting Cartesian cliché. The logic of this was probably not impeccable. Some have challenged the slipping in of the I. Perhaps 'thinking occurs' is all one can be sure of. But given the liberties Descartes then took in building up his system of reasoning (an optimistic proof of God's existence, leading into an arbitrary appeal to God's goodness, in order to establish the validity of both observation and reason) this particular point is not important.

Remember the problem Democritus had encountered two thousand years before with his atomic theory? If the world is mechanical, and thinking is real, what is the relationship between thinking, this immaterial conscious stuff, and the observed material world? And if the material world is determined, but the mental world is not (for we have free will, don't we?) how do these worlds interact, and how is that not a contradiction?

Big questions, and Descartes' dualism was the big answer. Thinking stuff, the human soul essentially, is separate from material stuff, he thought. Thinking may happen in the brain, but thinking is separate from the physical working of the brain. He even suggested where in the brain the non-corporeal thinking stuff might reside. How the two interact remained, in Descartes' version, a big fat mystery, and for this reason it is usually today rejected as inadequate. (That said, we're yet to find a really satisfactory replacement, and there are those who still choose to argue that it may be impossible in principle to do so. My penultimate chapter looks at this in far more detail.)

To his great credit Descartes himself tormented over the inconsistencies. As an amusing example, he had a thing for cross-eyed girls, the lingering effect of a boyhood infatuation, but he found that with great will he could resist this strange attraction. He saw this as a case of the mind making its influence felt over the body, but he had no idea about how this might have happened. He just wasn't sure how a non-corpereal will could exert its influence over a physical brain. And it troubled him.

Some have picked up on Descartes' deep religious convictions, and suggested that the dualist model was a way of appeasing the forces of mysticism. I think this is unjustifiably cynical. After all, he was presented with a genuine problem, one for which an obvious, satisfying solution is yet to be framed. So it's not hard to see dualism as his best crack at an answer. What it did do was partition off the realm of the physical from any spiritual influence, and allowed science to proceed with an agnostic form of enquiry, seeking in nature the underlying laws which would explain all its apparent complexities. Descartes shared this fixation on simple laws with Bacon before him, and Newton, who followed.

The word genius is designed for people like Newton, born in the year of Galileo's death. A magnificent mathematician, he discovered calculus (as did Leibniz, it must be said) which like Descartes' geometry revolutionised mathematics. But Newton's greatest fame rests upon his apparent discovery of the very laws of nature Bacon and Descartes had dreamed of, even if his methodology was not particularly Baconian. Newton's laws of motion explained how forces cause objects to accelerate and decelerate, and on this he built a system of mechanics which we still teach in schools today. Just as profound was his explanation of gravity, which he modelled as a function of mass and distance. He suggested that the very movement of the planets was the result of the gravitational attraction of the relevant masses, ceaselessly changing their momentum as they orbited. He used this to devise equations to describe planetary motions, and their elegance, along with their apparent accuracy, was enough to convince the world he was onto something special. This in turn created an interesting challenge for the church, with its Platonic view of the perfect heavens entirely beyond the influence of shabby earthly influences like mass and gravity, a blow every bit as hefty as Galileo's (although ultimately not a patch on Darwin's).

If Bacon and Descartes were the prophets of the scientific revolution, Newton was its messiah. He genuinely appeared to have delivered the goods, to have used the rationalist methods of reason and observation to unlock Nature's secrets. History shows us a more complex picture of course. A deeply religious man, Newton was sure his laws of planetary motion proved the existence of a God, who was needed from time to time to rebalance the equations (later calculation errors explained this better). He also devoted much of his life to alchemy, and in this the much vaunted scientific method yielded more modest results. And as we now know, the laws he uncovered were only the start of an explanation reimagined by Einstein at the start of the twentieth century. Today, modern physics awaits its next revolution, frustrated at the inability to marry quantum theory with general relativity.

But all of this we can say with the benefit of hindsight. At the time it must have been hard not to believe that the startling results of the new science weren't ample proof of Descartes' assertion that God would not deceive a truly inquiring mind. And for those not yet on the scientific bandwagon, there was also technological advancement to consider. The iconic example is James Watt's steam engine at the beginning of the eighteenth century, which led directly to an industrial revolution. With exploration came new trade opportunities, new wealth, and exposure to new cultures and ideas.

The optimism of the times saw theorists apply the emerging scientific methods to more complex, living systems. Geologists were finding hints of the ancient nature of the earth, and fossil evidence suggested a weird, pre-human past, again bringing science into direct opposition with some of our oldest and most precious creation myths. Thinkers like Locke went even further, looking to apply the new rationalism to matters of the mind, and to ethical and political systems. Such was the state of play at the height of The Enlightenment. The pendulum was swinging back and it looked for all the world as if Plato's idealism would be replaced by a more earthly form of thinking. The measurers and tinkerers had proven their great worth in solid, practical terms, with their predictions and their inventions. When they told the people, 'This is what the world is really like: behold the lessons of nature,' it is not hard to see the attraction of their claim.

There are many more great scientific thinkers from the seventeenth and eighteenth centuries worth mentioning, but these few are sufficient to make the point. By the end of the period 'a Thales-like curiosity' was well and truly re-ignited. New technologies, new attitudes, historical contingency and the infectiousness of success had pulled the human gaze back to the details of the surrounding world. The movement of the stars, the circulation of the blood, the turning of the seasons, the rising and falling of the tides, the colours of the rainbow: these things were no longer just to be observed and wondered at. They could be explained, modelled, predicted and crucially, manipulated. Humanity was wading into a whole new ocean of understanding, reshaping the world as it went.

As we have seen, this remarkable progress was enough to shake our faith in the old institutions whose stories were being constantly updated and replaced by science. But that raises a crucial question. Where exactly does this new science get its authority from? Why, in other words, is the scientific method, as championed by Bacon and practised by the many geniuses who followed, so damned powerful?

For some this might seem a trivial question. It works, so who cares why? Pretty much the attitude I have towards my computer. I turn it on, it works (thank you, Apple) end of story. But if we stop to think about it, there is a very good reason indeed to ask the 'why' question. For while I don't know much about the inner workings of a modern Mac, I do have a rudimentary understanding of computers. And it is just that understanding which shapes the way I interact with it.

I understand the principle of an algorithm for example. I understand that the link between the commands I type and the actions the computer performs can be reduced to a predictable input/output network, albeit a massively complex one. And thanks to this knowledge, I do not attribute magical qualities to the machine. I don't get angry at it. I don't try to read its mood, interpret the emotional subtext of its communications with me, or start to sulk if it takes too long to boot up. I don't take it personally — most of the time.

Contrast this with someone who perhaps had spent the last thirty years out of circulation, suddenly coming into contact with their very first computer. How easy might it be to dupe them into believing that technology had advanced far beyond its current state? They might believe that the modern machine had indeed developed to the point of possessing a wilful personality. It could be a computer that crashed when it was feeling unloved, ignored your commands when it was tired, and on a good day knew just which song to choose from its random playlist to suit your mood. And it isn't impossible to imagine such a user beginning to confide in the machine. (Have you ever spoken to your cat like you believe it understands you?) Maybe they would even seek the computer's advice, or one day in a fit of murderous rage, unplug the thing and hurl it into the mini-skip.

Our understanding of why something works is crucial for defining the limits of its powers. I trust my computer to tell me what 3 divided by 17 is, but not to tell me how this piece of writing makes it feel. ('Like a machine, damn you. You make me feel like I'm just a machine!')

As it is with computers, so it is with science. We must both make use of its powers and also define its limits. To understand its limits, we need a working understanding of how it manages its remarkable tricks in the first place.

To end this chapter I want to look at two of my favourite philosophers from the eighteenth century — who asked pesky and interesting questions about the nature of scientific knowledge, and in many ways first pointed us towards the modern outlook — David Hume and Immanuel Kant. Be warned, what follows is the densest part of the book, a brief detour down the tangled walkways of pointyhead philosophical hair splitting. If the very thought of that makes your skin crawl, there will be no harm done if you skip to the beginning of the next chapter. (Apart from missing out on something which, with a little effort, you might just end up finding quite irresistible.)

David Hume (1711–1776) was Scottish, a fact which I find inexplicably important, and so I mention it. He was a sceptic in the true sense of the world, not just shining his unbelieving light upon the old religions (he wrote a treatise on the non-existence of God, taking the precaution of not having it published until after his death) but also on the rise of the new scientism. Hume's great asset and burden was his curiosity. But unlike Descartes, he lacked the buffer of deep religious conviction. Just the sort then to raise his hand at the back of the classroom and say, 'But sir, there's still one thing I don't quite understand.'

Hume's challenge to the scientific method is a funny old beast. Some have seen it as one of the most important and pressing problems in the philosophy of science, while to others it is a silly distraction of no practical importance. Amongst those who took it seriously were some of the greats of western philosophy. Kant claimed Hume 'has awoken me from my dogmatic slumber'. The twentieth century logician Bertrand Russell considered Hume's case unanswerable, and considered this to be a genuine problem, while Karl Popper, one of the best known philosophers of science, devoted a great deal of his career to finding a way around it. So while you may be tempted to dismiss it at first glance, it's probably worth a second look.

The scientific method, remember, is based in part upon inductive reasoning. We make a local observation and use this to draw conclusions about the wider world. Or in other words, we assume the universe is subject to regular laws. Without this assumption, science can not move from observation to model. Taking such regularity for granted is the most natural assumption to make, flowing easily from the way we all learn. We go out in the rain, we get wet. We assume rain goes with getting wet. In fact we go further and assume the rain causes the subsequent wetness. It's a handy rule. It stops us from repeating the experiment every time the rain falls and means that at worst we will be burnt by fire or zapped by an electric fence only once. Animals can be trained by a similar principle. Only the wilfully stupid or philosophically obstinate resist its logic.

Hume was obstinate. Okay, he said, I accept we believe the sun will rise in the east tomorrow because up until now it always has, and so it's natural to keep betting that way. But isn't that exactly the point? It's natural. An instinct. The way fearing death or believing in the soul is natural. A product of our nature, not our reason. We believe the world is consistent because we have an instinct to believe such things. It's part of the human design. (See how closely the eighteenth-century atheist predicts the twentieth-century arguments of sociobiology.)

Induction, or consistency, is a tremendously useful, efficient assumption. When Eratosthenes found evidence the world between Alexandria and Aswan was bulged outwards, he assumed this bulging continued far beyond his observations and made the whole world a globe. But how rational was this step? It worked out pretty well, but was the method strictly speaking justifiable? Why didn't he feel compelled to check other bits of the globe?

Hume was essentially challenging us to show why all the evidence of regularity we have collected so far does not equally support a world where all the rules hold consistently — until tomorrow. Why couldn't our world be one where everything is the same, and then wham, one day it's all changed? Nothing that has happened so far makes change less plausible than the 'constant forever' thesis. Or how about believing none of the rules work consistently, except in the case where measurements are made, because we are being deliberately deceived by a mischievous creator who wishes to keep us from the truth? Or what if there are random non-uniformities which up until now random chance has stopped our measurements from picking up? Or what if the world is entirely consistent except when some unknown outside force chooses to jump in and muddle things around a bit, taking care never to be noticed? Logically speaking none of these options seems better than the other, and this problem has become what is known as the challenge to induction.

And yes, it is extremely tempting to dismiss the challenge as a load of abstract old nonsense. Why would we care? Assuming the world is regular allows us to build models of the world very efficiently, and what's more these models have proven themselves to be tremendously effective. Induction allows scientists to do their work, and science is very useful, so let's just leave it at that. Or as Bertrand Russell put it:

> What these arguments prove — and I do not think the proof can be controverted — is, that induction is an independent logical principle, incapable of being inferred either from experience or from other logical principles, and that without the principle science is impossible.

By assuming some sort of consistency in the world we are able to build up our expectations, assign probabilities and prepare for the future. Should we really care if the result is in some waffly sense philosophically suspect?

Well, that rather depends upon what you want to use science for. If you accept science is just an efficient method of building palatable models, a neutral data-cranking machine, then there is no problem at all. Who cares if the world might be otherwise? So long as the predictions are good for now, let's go with them. But it is patently clear that not everybody thinks this way. Many hugely respected figures in modern science see science as offering something more than just predictions. They see it as yielding a true description of our universe, often in a vague and misty-eyed manner redolent of the Pythagoreans. These are exactly the people who are most easily lured into making the leap from a widely accepted model to an idiosyncratic interpretation without realising they have crossed the science/storytelling barrier. And they, to my mind, need to answer the induction challenge. For others, who have never seen science as anything but a technique through which nature can be modelled and manipulated, the problem is far less pressing. The divide between scientists who see themselves as technicians and those who see themselves as stalkers of the truth hinges on Hume's challenge.

Interestingly this split in the scientific community has venerable roots. Many place Copernicus (1473–1543) at the very start of the scientific revolution. Now take a look at the compromise the Lutheran Andreas Osiander (1498–1552) went for in his introduction to Copernicus' theory of the solar system. He suggested that it:

> not be seen as descriptions of reality, but rather as computational devices.

Interestingly, the Catholic church offered Galileo the same escape route, but he considered it beneath him. He was, I can't resist pointing out, leaning towards a sort of mysticism when he did this. So much for the iconic science versus faith example.

The pithiest working of Hume's legacy is the famous 'is does not imply ought'. Hume saw in the induction challenge the need to keep our physical models and our stories of meaning separate. I like to think that by reminding us of science's utilitarian roots, he stripped it of its philosophical pretensions. Having done so, he then turned his attention to ethics. If we can not reason our way towards our moral obligations he wondered, then from where do our notions of right and wrong spring? Again his atheism deprived him of an easy answer. He wasn't able to say 'God has told us what is right and wrong, and so ethics in the end is a doddle' (which perhaps would be an excellent solution if different gods didn't have the nasty habit of telling us different things). So instead Hume turned to instinct. We humans, he suggested, have the capacity for empathy. It is part of our make-up. Our ethical systems, our notions of right and wrong, stem from this fact about ourselves. Because we are able to extend our feelings about and concern for ourselves to others, we have a fighting chance of building up a rational set of rules by which a complex society can operate.

But why might empathy be part of our make-up? Despite having come perilously close to formulating the principles of natural selection himself, Hume didn't quite have the evolutionary theory needed to address that question, and so we can only marvel at his prescience and wonder what he might have made of biology's recent advances. That wondering will have to wait for another chapter. The point I most wanted to highlight is the way Hume gave us good reason to be cautious about science's ultimate reach, advice which by and large we have woven into the fabric of our modern society.

Now I want to turn to another deeply original thinker, who though impressed by Hume's challenge, wasn't quite convinced. Enter, stage left: Immanuel Kant (1724–1804).

Kant's objection may well have occurred to you already, or at least suggested itself in that familiar fuzzy way when something just doesn't seem quite right. It's all very well to say that we use inductive reasoning because it's so very useful, and to make the distinction between usefulness and truthfulness, but isn't that only telling half of the story? Isn't the next important question 'Why is induction so useful?' Why is treating the world as if it's subject to uniform rules such a powerful method of predicting the physical world's behaviour? Hume in essence had said 'because it just is' and that sort of answer is a red rag to the enquiring mind. Kant wanted to go further, and see if the reason that induction worked couldn't in turn tell us something important about the true nature of our universe.

One possible reason why induction works so well is the reason most scientists would probably give if pushed on the matter. Perhaps the world really is a regular beast. But Hume had already shut down that avenue, and Kant accepted his block was valid. It's certainly possible the world is regular, but it's only one possibility. So if by reality we mean something like 'most likely explanation' or even 'most palatable explanation' then that's fine, and this watered down version is pretty much the modern view. Kant took a slightly different, and impressively imaginative, line.

Kant argued that the uniformity we see doesn't happen in the world so much as it happens inside our brain. So while we can not be sure the world conforms to regularities, we can be sure our thinking does, and that sort of certainty is as good as we'll ever get. In his own words:

> The understanding does not draw its laws from nature, but prescribes them to nature.

Now that's the sort of statement that's not going to yield its meaning without a struggle, so let's see if an analogy might offer some clarity. Imagine you're dredging the ocean with a fishing net, in order to survey the sea's life forms. (This example comes courtesy of a physicist, I believe; physicists do great analogies.) You haul up the net, look at the little critters you've caught and say, 'Oh look, there's nothing in the ocean smaller than the mesh of my net, I've discovered a rule of nature.'

Only you really haven't. What you've discovered is a law of nets. The size of the critters reflects not the size of ocean life but rather the size of the measuring device. For Kant, our mind is the net. You can't find what you aren't capable of seeing, and at some level we are simply incapable of conceiving of disorder. So we must find order, or laws, in the things we examine.

But what sort of disorder was Kant talking about? Surely not the disorder of a school playground or a late night bar, for imagining these is no stretch at all. Kant was after something much more radical than this. What he meant by disorder was things which are outside of space, time and causation. These are the things which slip through our net.

Think about how challenging that is. Space, time and causation are not thought by Kant to be facts of the physical world so much as they are facts of the mental world. We can't think of an event which happens at no particular time, in no particular place, for no particular reason. Our brains can't cope with that. So in order for us to understand it, we describe to ourselves anything that does happen in terms of space and time and causation. Events as we perceive them happen in a given place, at a given time, as the inevitable result of a given set of circumstances. That doesn't mean that time, space and causation are what the universe is made of, they're just the language we translate our experience of the world into, in order to think about it. And the language we translate into is an ordered, regular one, so it creates the type of order in our measurements and models that serves inductive assumptions so well.

To try another slightly strained analogy, consider a newspaper report of a football match, unaccompanied by any photographs. Imagine too a person whose only experience of football is reading such reports. The real match has many qualities (sounds, smells, movements) which the type-set report can only hint at, and which the reader will never properly appreciate. And in the same way the reports have certain qualities (structure, brevity, visual regularity) which are not qualities of the match itself. There is something of the flavour of this in the way Kant suggests we make sense of our world. We experience and interact with a representation of the world, but not the world itself. And this representation Kant thought, might explain why induction works so well. Hume was right. The world may or may not be regular. But our perceptions of it are, and as this is all we'll ever know, it's all that really matters.

We could waste a few more trees considering now how well Kant answered Hume, and we wouldn't be the first. ('But why are our brains this way?' is the obvious starting question: feel free to continue on from there.) I want to move on, and so will only note these two things. First, in terms of the science/storytelling divide, Kant, like Hume, separated ethics from the physical models of the universe we build. He doubted that the way we perceive the world could account for the 'moral certainty within'. He then proceeded, with a mix of rationalism and religious vigour, to examine the puzzling phenomenon of our sense of right and wrong. His term for the deeper unknowable realities, the transcendental, was later hijacked by romantics and hippies, and so were launched a thousand house buses.

The second thing worth noting is how impressively modern Kant's view now feels, as if he had somehow anticipated the path modern physics would take. Today many of our physical models ask us to employ mathematics which makes little intuitive sense. Try to get a physicist to explain quantum mechanics, string theory or even general relativity, and you will soon see just how lost for satisfactory metaphors they have become. In quantum mechanics we are asked not to think of particles having a set position in space, but rather as having their occupancy of space described by a spreading out probability function. One current attempt to formulate a solution to the quantum gravity problem suggests the world is best described as being two dimensional, with the third dimension and time itself emerging as a sort of illusion. This sounds all very sci-fi whacky, but what if such models prove to be better predictors of physical effects than our more intuitively satisfying time and place descriptions of reality? Which then deserves the title true?

Predictably enough mystics are scrambling to fill in the gaps. ('See man, it all means there's infinite universes and we exist in all of them, and every time a quantum state is settled the universe splits again … pass the bong will ya.') Kant may provide a better explanation. The world as it is, and the world as we are able to conceive it, may be very different things indeed. That this difference is bridged by the abstractions of mathematics would surely have warmed Pythagorus to the core.

Okay, that's enough of this hair splitting business. Before leaving this chapter, let's quickly look at where we've got to.

The rise of reason, as typified by Thales' early exploits, gave us a new way of looking at the world. Rather than just listening to the stories handed down to us, we could turn them over in our minds, compare them to the things we had experienced, and ultimately question their conclusions. This spirit of curiosity led to great advances, but the price we paid was a level of uncertainty that few feel comfortable with. Plato then was the great analgesic, assuring us that even within a world of reason and argument, we could rely upon the discovery of greater truths, made available to us by some shadowy, mystical means.

It was this brand of mysticism which the Christian church inherited, and within the Western tradition truth by revelation dominated the landscape for another two thousand years. A confluence of events led to a new interest in the methodology of close observation and experimentation, and the great success of this new science cast doubt upon the power of the church. As the church's technical prowess was challenged, so too was its metaphysical worth. And into this gap science emerged as a contender, a new bringer of truth.

But truth for a human being is about more than forces and chemical reactions. We also seek a place in the universe, a perspective from which we can gain comfort and motivation. We seek an understanding of ourselves and others which allows our interactions to be stimulating and satisfying. In short, we seek meaning. Without this we have no motivation, no notion of others, no ability to interact. Reduce it all to physical prediction, and the self — the reason for living — disappears. By the eighteenth century some great thinkers were beginning to cast doubt on the ability of scientific models to provide this sort of 'truth-model'.

In the modern world we use the institutions of science to supplement our day to day experiences, at least when it comes to building models. This model building is what we often mean when we use the word truth, and in this commendably restrained sense, truth does indeed exist. To say otherwise is to deny common usage (although a lot of confusion would be avoided if we found a new word for it). It is, in a simple and non-controversial respect, quite true that the sun is hotter than the earth and that Tokyo has more people living in it than Eketahuna. In day to day terms these are what we think of as facts about the world and we are unlikely to get into any sort of trouble regarding them as such. (Of course, if it turns out that models which treat aspects of time and space as illusory are better prediction generators, then even these apparently simple truths must come with caveats.)

When it comes to interpretation though, things get decidedly murkier. For the word truth is also commonly used to refer to shared beliefs. We might want to claim that the revulsion we feel when a young child is beaten to death also reflects some deeper moral truth about the world. But any consensus is easily strained when we look for a sensible starting point for the definition of a child: witness the tortuous debates on abortion, stem cell research and reproductive technologies. So how on earth does a society go about forging agreement in these areas?

Traditionally some form of religion played an important role here. Deeper truths were made available to us by way of contemplation, revelation and instruction. A particular definition of life, for instance, was made solid by the certainty that this too was God's definition. Religious traditions though have been dealt vigourous blows by our intellectual progress. On the one hand scientific discoveries (cosmology, evolution, cognitive psychology, to name a few) keep undermining the foundations upon which our religious stories are built, and it gets harder and harder to keep the faith. On the other hand a shrinking globe reacquaints us with Protagoras' famous observation: 'The fire burns the same way everywhere, but the law of the land differs from place to place.' Different groups have vastly different stories to tell about the deeper truths, and if they're all 'right' in some sense, this rather diminishes what we mean by 'right' in the first place.

In the modern world, we are in danger of being lost, disconnected from our stories. One possible solution is the optimistic hope that the scientific method itself might happen upon the 'true stories', those immune to continuous revision or regional variation. It is natural to want science to be able to do this job; it would just be such a wonderful outcome. So, often unwittingly, we attach quaint little stories to our scientific models and hope nobody will notice.

Those who cling to this hope reject the challenges of Hume and Kant as inadequate. Surely, they reason, the great success of the scientific method must tell us about something more than the nature of the human mind. Doesn't it also tell us something about the real nature of the universe? Doesn't it suggest that science is somehow edging its way towards a deeper truth? Karl Popper, said to be the philosopher most read by scientists (or the only philosopher read by scientists) thought so; and so now we must think about Karl Popper.

3 Popping Popper?

Let's indulge in a little fantasy. Imagine we could send two people back in time, from the present day to the birth of the scientific revolution: the sixteen hundreds. Let's say one is an expert in the history of modern science, the other equally well informed on advances in philosophy and the arts. Now consider the wonderment as the futuristic road show makes its way through the century. Galileo, Bacon and Newton shuffle forward to the time machine, and are told strange scientific tales from our future land. What would they make of the theories of relativity, I wonder, or the structure of the atom, nuclear power, evolution, combustion engines, genetics, quantum mechanics or computers? It would take many years to adequately report to them on the advances, and if they were armed with such knowledge, we can be fairly certain they would change the course of history (who knows, perhaps even destroying the planet before the time machine in question could be invented, so plunging us straight into one of those annoying time travel paradoxes).

Such logical impossibilities notwithstanding, our science expert is kept tremendously busy, and no doubt is treated like some sort of demigod. But what's our arts and philosophy specialist up to? Poor Descartes is beside himself with excitement. He can't wait to hear just how his mind/body problem has been resolved. Hume, always with a twinkle in his eye, is curious about how the induction problem turned out, and what the current thinking on God might be. And

Shakespeare wants to know about the breakthroughs in theatre that have since dwarfed his own humble contribution. But here the story takes a strange twist. Our unfortunate visitor from the future is going to be a little less impressive than her colleague. For the mind/body problem is still a problem, the inductive challenge still a conundrum, religious tensions appear as far as ever from being resolved, and in schools today, when we wish to demonstrate theatrical excellence, Shakespeare is our example.

Okay, it's just a little fantasy, and it's easy to manipulate such things in order to make a point, but nevertheless there's something important here which needs to be accounted for. In the last chapter I rejected the notion of science as a mystical path to deeper truths. Science's appeal is more pragmatic; we rely on its methods for no other reason than they have worked well in the past. It provides us with knowledge which is probabilistic rather than absolute. But that begs an important question. What is it exactly about the method that makes it so successful, and accounts for its apparently progressive nature? Why does this business of observation, prediction, experimentation and generalisation work? 'If I have seen further,' Newton wrote, 'it is by standing on the shoulders of giants.' And so even the most workmanlike scientist today can achieve feats Newton could never have even dreamed of. Meanwhile, down at the local repertory, well they're not exactly making Shakespeare redundant are they? So how come the difference? If we can't solve this puzzle, then perhaps we have no right to insist science stay in its model building box.

So in this chapter I am going to turn again to another of history's big brains, a man who gave this question considerable consideration. From whence, Karl Popper wondered, does science's curious power derive? Science is plainly different from other disciplines in respect to the predictions of its models and the technology these predictions enable. Satellite technology makes GPS extraordinarily accurate, cellphones send photographs around the globe in the blink of an eye, organ transplants are now routine and planes really do stay in the sky most of the time. As much as story and cultural construction are important to science, that can't be all that there is to it. Show me a post-modernist at thirty thousand feet, goes the old quip, and I'll show you a hypocrite. Something has to explain technology's advance.

Popper (1902–1994) was wary of falling back on the easy, mystical answers. For him it was not enough to assume that science had simply, through good luck or divine guidance, stumbled upon the truth of the universe. He was well aware of Hume's induction challenge, and believed it to be a significant hurdle in need of negotiation. Nor was he prepared to lapse into the wholly utilitarian response of 'Look, it works, that's all that matters, why question it?' For he was a philosopher, and philosophers don't tend to think of 'It just does, okay,' as an excellent answer.

In terms of science, Popper's challenge was twofold. He first sought a defining characteristic which could sort the scientific from the non-scientific: what came to be known as the demarcation problem. And from here he sought to explain why, despite Hume's warnings, the method seemed to work so very well: what we think of as the induction problem. So, how did he go?

Popper regarded himself as a common sense philosopher, in that he believed philosophy should deal with practical issues. He saw philosophy as the leading edge of discovery, defining the ground upon which the practical sciences could establish their basis of inquiry. I have already used two of his favourite examples when it came to making this point. Plato was in part trying to work through the implications of the discovery that numbers could be irrational, and so was attempting to establish the limits of rationalism in our understanding of the world. And Kant was specifically trying to answer Hume's problem of induction, so as to understand how it was science arrived at its special transformative power. Popper was dismissive of both their conclusions, but he saw them as good examples of the way one should proceed: i.e. defining a problem of practical importance and then attempting to clarify the issues involved.

One of the 'common sense' problems Popper dealt with was political. He railed against historicist thinking — the belief that human history is unfolding according to set rules which can be studied and, in the most dangerous interpretations, helped on their way. The idea that such progress could carry the moral weight of inevitability appalled Popper: hardly surprising given the world he had grown up in. The early part of the twentieth century was a case study in the excesses of visionary movements: consider Social Darwinism, the eugenics movement, the power propaganda of World War One, the justification totalitarian states found in the Marxist historical narrative, and ultimately the genocide of World War Two. Living through such times it must have been easy to believe the greatest threat to humanity was posed by those who believed they were serving history's greater purpose.

The phrase 'greater purpose' takes us straight back to the science debate, and Aristotle's lasting legacy. Is it within the capacity of the human intellect to ever discern a greater purpose, or is the very notion of purpose the preserve of the imagination? While some people today worry that letting go of absolute moral standards will lead to social decay, Popper lived in an age where the exact opposite seemed to be the case. Those who were most sure of what life was all about seemed capable of doing the most damage. Popper traced these dangerous trends in Western thinking back to Plato's idealism, and in particular his model Republic. Against this he proposed a political science of muddling through, of cautious tinkering and compromise. No grand plans, just dealing with problems as they arise, protecting people from deprivation and abuse, but leaving them free to construct their own stories, and find their own ways forward to happiness. It was a liberal call for tolerance based upon scepticism, and in this I share many of his prejudices.

Unsurprisingly, these same themes emerged when Popper turned his mind to science. Once again he sought to make a virtue of uncertainty. He explained his interest in the demarcation problem by tracing it back to the intellectual environment of his university years. In those days, Marx, Freud and Adler were all the rage. Popper recalls that he, like so many, was initially impressed by the remarkable explanatory power of this thinking, and the compelling certainty of its adherents. It seemed there was nothing that could not be explained by the dynamics of the class struggle, the tugging of the subconscious or the damage of the self esteem. This type of thinking could overnight turn a callow youth into a world weary expert on absolutely everything.

'Did you see that bus crash this morning?'
'Mmm,' replies the emerging Marxist. 'The driver was being exploited, more labour was being extracted from him than he was being paid for, and he came to see the machine he was driving was really driving him. He rebelled. Full respect to the comrade.'
'No way,' counters the youthful Freudian. 'He was driving past that bus stop, and his subconscious noticed the woman in the advertisement, who looked just like his mother in her wedding photo. His damaged id rose up, trying to assert itself, and the crash you witnessed is nothing but a by-product of the ongoing battle for his psyche.'
'Nonsense,' the adolescent Adlerite contributes. 'The bus driver was the second child of four, and when he was growing up his needs were often overlooked. His self esteem was damaged, and when the woman at the last stop didn't return his smile, it was one rejection too many. He was trying to kill himself.'

All it takes is a little theory and a big imagination and there's nothing that can't be made sense of. And therein, Popper realised, the problem lay. If a theory is so flexible that it can happily accommodate any occurrence or observation, then what can it actually tell us about the world? Certainly it will be lousy at making predictions, for all outcomes are supported by the theory, and therefore none are demanded of it. The biologist David Penny likes to tell a story about a little boy who is a perfect marksman. On his target range wall, every single slug shot has passed exactly through the bullseye. 'So how did you achieve such accuracy?' someone asks. 'Easy,' the boy replies. 'Shoot first, then paint the bullseye.'

This sort of post-hoc omniscience creates a problem of practical importance. For while there may be a certain recreational value to be found in constantly reinterpreting the past, the problem of how to avoid future bus crashes is not well served by it. Someone really ought to check the brakes.

Lest you should feel I am being unfair in lampooning the social sciences in this way, let me assure you in advance that Popper was well aware that scientists too could fall into exactly the same trap. Indeed we can imagine a budding evolutionist proudly proclaiming the bus crashed because: 'the driver's reactions were not sufficiently sharp, and so he was selected out of the population in favour of drivers better suited to their environment. It's natural selection.'

However, Popper felt that the strength of science derived precisely from its ability to avoid this type of error. He didn't claim all science would; only that good science should. And on the back of this distinction, he built his model of scientific discovery.

We have already come across many of the characteristics of the scientific method. Careful, verifiable observation. Deductive logic, moving from what we see, to what it might imply. Careful experimentation, to test these implications, and inductive reasoning, taking the observed relations and generalising them. And of course the odd moment of inspired hypothesis-forming lunacy, without which the other techniques are rendered impotent. (Perhaps it's all water!) Amongst all of this Popper was seeking to find that quality which could make science stand apart, and so explain its impressive progress.

The characteristic he went for is a deceptively simple one, although like so many simple ideas, prone to much misunderstanding. Science, he concluded, is falsifiable. What he meant by this is straightforward. If I make a statement about, let's say, the cause of a bus crash, then it is scientific if I can imagine a piece of data which would force me to change my point of view. 'It was the brakes' by this measure, is in some loose sense scientific. If the brakes turn out to have been functioning perfectly well, then they can be ruled out of the enquiry. In fact they must be ruled out. But how about 'it was because he was exploited as a worker'. 'No it wasn't,' you point out. 'The brakes failed.'

'Yes, but if he hadn't been exploited as a worker, he wouldn't have been put in a bus with bad brakes,' comes the reply. 'It's capitalism, man. Capitalism is killing our bus service.'

The difference is in part attitudinal. The budding Marxist just doesn't want to let it go. But there's another difference too, the difference Popper is getting at. Before we attend the scene, we can clearly define what would or would not constitute sufficient braking capacity (vehicle inspection standards do just this) and hence say, 'If the brakes meet this standard, we will not consider bad brakes to be the cause.' Can we make a similar statement regarding the exploitation of the working class? At precisely what level of equality, pay, training, staff morale, would we agree to let go of our theory?

As I said, this simple idea has been subject to much misinterpretation, some of it wilful, so I'm going to spend a little bit of time fleshing out both the theory and its implications before considering its biggest weaknesses. An iconic example of this method of falsifiability in practice, and one much used by Popper, is The General Theory of Relativity, posited by Albert Einstein (1879–1955). One of the implications of Einstein's theory was that light would be bent by the gravitational pull of large objects as it passed close to them. This provided a way of checking the theory, as long as two provisos held.

The first proviso is obvious but is sometimes overlooked. An experiment for testing the bending of light must be feasible. In this case the distance had to be great, and the mass significant: the test couldn't be put together in a lab with a torch and a ball bearing. The plan was to measure whether the light from stars would be distorted as it passed close to the sun. In this situation another problem raises its head. How will we see light that passes by the sun? Won't it be drowned out by the sun's light? (We don't see stars in the daytime sky.) The solution proposed was clever. Wait for an eclipse, and measure the apparent relative positions of stars close to the sun at that time. Compare this to their relative position at night, and use this to see if any distortion has occurred. So it could be done. At this point, we can say the theory is falsifiable. If we think there probably will be a way of testing a theory, but right now we're damned if we can think of it, then we're not yet in a position to call it scientific, at least not by Popper's definition.

The second proviso is more interesting. Popper developed the notion of information content. The experiment should test for a result that isn't already known to us. Why? Because if the result is already known, then it shouldn't be regarded as a good corroboration of the theory, simply because the theory was drawn up with that information in mind. Remember we're trying to get away from the dangerous habit of subjectivity, of manipulating story and data until they fit together in a way that supports our preconceptions. An example might clarify this.

A person notices that men are often taller than their female partners. 'Hmmm,' says the researcher. 'I propose this is because in our evolutionary past, men with shorter partners were able to hold them close while simultaneously looking out over their heads for signs of danger. A great advantage in that competitive environment.'

'Excellent,' you reply. 'Nice story. But it isn't scientific because it can't be falsified.'

'Of course it can,' comes the mistaken reply. 'If men are on average shorter than their partners my theory can not possibly be true. Furthermore, if it turns out tall men can not see over shorter women's heads, again I would have to admit my theory is falsified.'

The problem here is apparent. You cannot use the information you used to propose your theory (men are taller than women and can see over their heads) to also verify your theory. It's circular. It's also disturbingly common. In the field of evolution such reasoning is often derided under the title, *Just So Stories* after Kipling. The thing that impresses us about the Relativity example is that, prior to Einstein, nobody had any reason to expect the light bending effect. So, when the measurements were taken and the effect showed up as predicted, people had good cause to be impressed. This is what Popper referred to as information content. Think back to Eratosthenes. No one would have predicted that the angles of downward pointing wells were not parallel without the round earth hypothesis. Likewise, Newton showed that a rainbow of colours could be reassembled into white light, a result only predicted because of his hypothesis that the rainbow effect was caused by the splitting of light into its constituent parts.

So it looked as if Popper might have been on to something. Here was an apparently simple criteria that could be applied to any statement to ascertain whether it should be considered scientific. What is the novel prediction your model makes, and can this prediction be verified? If no such prediction can be found, then the model is interpretative, rather than predictive, and we should not consider it scientific. By this definition, if I tell you standing in the rain makes you wet, I am being scientific. You can test this out by standing in the rain and observing the result. If we wonder whether it is the new cat food which is causing Moggy to be so hyperactive and furniture unfriendly, we are also on the path towards science. The prediction is that a change back to the regular food will alter the behaviour. If this does not happen, we might consider the food hypothesis falsified. By contrast, a statement like 'Annie is competitive because she's still subconsciously trying to win her dead father's approval' is harder to test, because it's not clear what predictions it is making. It is an interpretative story. Note that we are not saying that such interpretations can't be good fun, and indeed enlightening, but Popper proposed that they should be put in a separate category from predictive, scientific thinking.

But why bother which such a distinction? Here Popper's agenda grows grander, and weaknesses appear.

Popper wanted to show how the demarcation between science and non-science could explain science's progressive nature. Through doing this he believed he had also answered Hume, by showing that we do have cause to believe science is approaching the truth after all. Popper's thinking went something like this:

Because scientific statements are falsifiable, we have an automatic mechanism through which progress can occur. The television reception goes fuzzy. We wonder whether the aerial may have been loosened in last night's storm. The prediction is that a trip up on to the roof will uncover a compromised aerial. We clamber up and find no such thing: the aerial as far as we can tell is fine. This allows us to move on, checking the attachment at the back of the set, and here discovering a loose connection.

Falsification, the non-match between a prediction and a result, provides a potential mechanism for moving forward in our understanding, and this is exactly the common sense progressive characteristic Popper was seeking to explain.

With regard to Hume, the point Popper seems to be making is something like this:

Hume has shown us that we have no good reason for believing what is true for the part must be true for the whole. Noting that all the horses we have ever seen are black does not guarantee all horses are black. But, once we have happened upon a contradiction, this contradiction does hold always and forever; once we have seen one brown horse, we can safely conclude that not all horses are black. We now know a general fact about the world, albeit a negative one, without having to rely upon induction.

From this Popper took it that science could proceed not by discovering true statements, but by slowly contradicting false ones, and replacing them with better, not-yet-falsified theories. In this way, Popper suggested that although science might not uncover absolute truths, it could slowly approach them. For example, Newton was not exactly right. We know this because Einstein showed inaccuracies in his predictions, and provided another model which explained them better. In time, another great mind may well push the Theory of Relativity down the pecking order. With each new theory, a little more is explained, because the new theory must always be able to deal with the data which falsified its predecessor.

The essence of Popper's definition of science is that it is a procedure which follows a method of proposing hypotheses and then attempting to show they are wrong. His contention was that this is the most reliable way of uncovering the best models of the relationships we observe in nature: models with the highest explanatory and predictive power.

This fits in nicely with our definition of science from the last chapter, where we saw it as providing not certain, but rather probabilistic knowledge. The more often a model has provided accurate predictions, the more attempts at falsification it has resisted, the more confident we are of its reliability. We believe the earth is roundish not just because of Eratosthenes, but also because of successful circumnavigations, and the view from a spaceship window. It's as close to a fact as we might want, and few would bet on it being revisited in the foreseeable future.

So we might argue that from Popper's point of view, the great technological advances, and the grand narratives of science upon which they were built, has emerged as a result of a certain attitude towards knowledge. This is an attitude of careful definition, clear reasoning, controlled experimentation and vigourous attempts to falsify. He was not arguing that all of science would meet these standards, but he was arguing that all science should aspire to it. Implicit in his argument is the belief that science makes better advances under this attitude than any other.

There are two aspects then to Popper's formulation, and it's worth keeping the distinction clear in our heads. By explaining progress, Popper was taking a utilitarian stance. We promote falsification because it works so well, and it works so well because it provides a mechanism through which the value of models can be compared. A model which has not yet been falsified is considered better than one that has, for it yields by this definition a greater range of reliable predictions, and prediction is the thing we are trying to achieve. Because we have a clear means by which arguments can be arbitrated, in science it is possible to 'move on'.

By attempting to answer Hume, Popper was also reintroducing the concept of truth into the equation. We can't know if we have arrived upon the truth of the matter, Popper suggested, for a not-yet falsified theory may well be falsified tomorrow. But by following the scientific method, we can nevertheless approach the truth, by discarding the statements which have been shown to be not true. With this talk of truth, just the slightest whiff of Platonic idealism wafts back into the house. Perhaps if we are approaching truth, science isn't just providing us with useful tools, but also with lessons about how we should think about the world…

So how successful was Popper in establishing his line of demarcation, and in answering Hume? If we wish to critique Popper (and many have) it seems to me there are three approaches we could take. The first is to say, 'That would be a good way of doing science, but because no one really does science this way, it's a bad metaphor for scientific knowledge. It's misleading, and gives science a credibility it does not deserve.' The second, and I think more interesting, is to say, 'Actually, that's not the best way of doing science at all. Focussing on falsifiability will not lead us to the most consistent and useful models, and here's why.' The third, and most oblique, is to argue, 'You do half the job well, and the other half poorly. Your model explains progress well, but only in utilitarian terms. The talk of approaching truth is misleading.'

I'll take you through these in order, dwelling in particular on the important contribution to the debate made by Thomas Kuhn, and then at the end we can consider which of Popper's insights are still standing. My personal opinion is that the last attack is the easiest to sustain. The first, however, satisfying as it does the tribal urges of academics, has had the widest airing.

How well does Popper's definition actually match what is happening in the real world of scientific research? Is science committed to falsification? To answer this we should be aware of a sneaky trick that is often pulled when describing the falsifiability model. We tend to use examples of successful science — that is science that has not been falsified — in order to explain the process, e.g. The General Theory of Relativity. We show how the theory *might* have been disproved; only the results came through, so everybody was happy. We offer up a description of falsification without, well, the falsification. But what happens if the data does not confirm the hypothesis? Is it really as simple as dismissing the hypothesis and moving on?

Let's look at an example. Little Maximillian is a budding scientist. He's watched bees flying around his garden, visiting the flowers. He's a quiet, studious boy and has read all about pollination, and his mother keeps a bee hive, so he's both interested and reasonably well informed. He's sitting on his porch one spring, and is taken by the sweet perfume emanating from the flowers. Perhaps, he thinks to himself, the reason flowers smell so sweet is that bees use the scent to find their way to the nectar. So there we have it, observation, leading to hypothesis.

Now Maximillian must use his nascent deductive talents to devise an experiment which might test his hypothesis. *Here is how*, he reasons. *If I can make a flower which does not usually attract bees smell like one that does, then the bees should start coming to it too, if my hypothesis is correct. If they don't, I will have falsified my hypothesis.*

He knows just what he will do. He takes his mother's perfume, which smells damned close to flowers, he thinks, and sprays it on some plastic flowers he was sent by his auntie who lives in Australia and still thinks he is a girl. Never mind. He plants the sprayed flowers in the garden and watches. The bees ignore it. He is crestfallen. His hypothesis was wrong.

Or was it? Maximillian starts to think a little harder. *Perhaps it is not the hypothesis that is to blame,* he thinks, *perhaps it's my deduction. Perhaps, despite the fact that bees use scent to find flowers, they use it in combination with other clues: sight, texture etc. My experiment would not have picked this up. What I should have done, is found a way of knocking the scent out of the attractive flowers. My reasoning was just poor. Or, alternatively, perhaps it was the experiment design. Perhaps I put the plastic flowers in the wrong place. Or perhaps that perfume doesn't smell like flowers after all.*

That's a trite example, but you get the point, I hope. You're never testing an hypothesis alone in science, you are also testing your deductive links from hypothesis to experiment, and your experiment design. Hence, there is never any way of being sure whether a falsification shows a bad hypothesis, bad reasoning, or bad experiment design. And even when there is good cause to doubt the hypothesis, often the experimental results can be made to fit a slightly modified version (bees are attracted by scent except when the flowers are plastic!).

This gives Popper's critics a promising foothold. For it would appear that falsification, which is fine and dandy in theory, in practice has no set criterion. When the results are positive, we assume the experiment was excellent. When we don't get what we were after, we make excuses, adjust the experiment, play around with our assumptions, make sneaky little ad hoc changes to the original hypothesis. Because there is so much wriggle room (normally the term used is wiggle room, but for me wriggle better reflects the sort of squirming going on) without any objective measurement of appropriate wriggling techniques, the decision on whether or not to reject a theory is by necessity a judgement call, and that makes it sound suspiciously like every other discipline. (Even the iconic Relativity example did in fact involve heated debate about the experimental technique and the data's interpretation.) The line of demarcation then is illusory.

Do I buy this attack? Yes and no. The reason I'm only partially moved by it is because it's a sort of argument by degree: 'because we can't show that science is perfectly objective, we must dismiss the value of objectivity in science'. Popper was trying to show that this is the approach we should aim for. Yes, we can be too reluctant to let our hypotheses go, and so reason poorly, blaming experimental conditions and making ad hoc adjustments to our theories. But we shouldn't: just as we shouldn't invent data for the sake of support. The fact that sometimes, in all disciplines, people fabricate whole studies does not in itself invalidate academic enquiry.

By the same token, the fact that there is always wriggle room for the poor practitioner to exploit only means that we must be as vigilant as we can against poor practitioners. This is why there is an insistence on reproducible results, and why papers are subject to peer review. It is also why a certain gladiatorial spirit is encouraged within academia, to catch out the sloppy thinkers. So we can reduce this argument to a reminder that science often falls short of its ideals, and to a reminder that we should judge science by the methods it employs, rather than the names it gives itself. Popper himself accepted this, and indeed based much of his criticism of various fields of intellectual endeavour on precisely this point. The point which stands, despite this attack, is that in order for science to progress, it must be committed to the process of falsification. Without it, we are in danger of chasing our tails.

There is another way of looking at this criticism. By highlighting the importance of wriggle room, it shows the difficulty of creating criteria according to which science can be tightly defined. Given this was one of the main problems Popper was trying to solve, we shouldn't ignore it. Science excels at the simple stuff. Think back to Eratosthenes: a well, some sunshine and a weighted string. That's a pretty simple experiment. There aren't so many variables. If the result isn't the one we're expecting, the wriggle room is minimal. When we're getting to little creatures like bees, there are many more variables to control for. Maybe it was the wind. Maybe the time of day. Maybe they were just returning to the flowers they'd visited that morning. As the system under study becomes more complex, the idea of falsification becomes far less clean. It becomes hard to lay out in advance what will or won't falsify an hypothesis.

So we are forced to make statements like, 'We expect to see some increase in immune system function', or 'We expect to see similarities in DNA sequences in the two genes being studied', or 'We expect to see that male aggression varies in response to environmental factors'. But how much increase in immune function will be significant, how much similarity do we need between sequences, how will aggression be measured? In such cases the deductive steps from hypothesis to experiment get quickly mired down in real world variables which can neither be tightly controlled nor modelled. And biology is as nothing compared to the complexity, say, of an economic system, where the variables being considered are the behaviours of human beings. In turn we could perhaps argue that predicting interest rates is simpler than predicting whether Jane and Malcolm's marriage is going to last.

It is no coincidence that the best examples of falsification experiments come from physics. Take for instance the prediction by the Irish physicist John Bell (1928–1990) that if two entangled subatomic particles do in fact communicate remotely with one another, in the sense that observing one will automatically effect the state of the other unobserved particle, then we should expect to see certain rates of coincidence when we independently measure aspects of their spin. Without getting sidetracked by the details, the important point is that at the time (the 1960s) this was not a popular hypothesis. It challenged too many long standing tenets; there isn't supposed to be action at a distance like this. Einstein himself had dismissed the result as 'spooky'. Yet here was an experiment which could be carried out in such a way that all observers would agree on the significance of their results, regardless of their prior investment in the outcome.

History shows the results were every bit as spooky as quantum physicists had predicted, and despite their counterintuitive nature, people were forced to accept them. These days ingenious new encoding systems are being built around this very discovery. In the relatively simple world of controlled physical experiments, falsification works well, and adds an important string to the scientist's bow. But as we move into the study of more complex systems, as we must, falsification simply becomes less and less helpful. We can test the 'Malcolm and Jane will stay in love' hypothesis, but only by waiting it out, and by the time the verdict is in, the information is clearly useless in a predictive sense (but brilliant for those who love nothing better than saying 'I told you so').

The more a discipline lends itself to tightly controlled experiments, the more powerful (in terms of prediction) its models become. We need not be too concerned that the definition of tightly controlled is a little fuzzy. After all, the line of demarcation between yellow and orange is also a little fuzzy, but this doesn't stop the concept 'yellow' being tremendously useful. We are saying that the easier it is for impartial observers to agree upon criteria for success or failure, the more likely we are to see disputes resolved, and to observe measurable progress. By implication, in all disciplines we should attempt to control our experiments as tightly as possible (through careful definitions, setting out in advance our criteria for falsification and so forth). This delivers us up at the doorstep of the second line of criticism. What if this last link in the argument is where the weakness lies? What if there's a better way of describing science's progress?

Here the oft cited philosopher and historian Thomas Kuhn (1922–1996) enters the fray. I approach this dispute with a certain caution, for it seems to me the Popper vs Kuhn debate has been greatly overhyped and typically generates more heat than light.

Kuhn approached the question of scientific progress from an historical perspective. Many people since have seen his work as standing in direct contradiction to Popper's, and in some circles this has morphed into one of those frustrating knee-jerk responses: 'That's not how we see science any more. Haven't you read Kuhn?' By this they mean his seminal *The Structure of Scientific Revolutions*. One of the reasons I approach this with caution — and perhaps here I am missing something — is that, when I read it, I thought the two theories were quite complementary. Let me just outline very briefly what Kuhn was saying, and then we'll see if he and Popper can't be, well, synthesised.

Despite subsequent publicity, the aim of Kuhn's work was not to refute the falsification hypothesis so much as to attack the notion of gradual and incremental progress in science. Kuhn was prompted to write his book, he says, because he couldn't fit his own reading of scientific history against the dominant story of scientists slowly improving upon one another's models, piece by piece. Rather, as the title of his book suggests, he saw science as lurching from one revolution to another. The conception works a little like this:

Ideas don't occur in a vacuum. The very questions we ask are constrained by our existing knowledge. Traditionally, students of science are brought up to believe in a dominant set of facts and assumptions: what Kuhn refers to as a paradigm. This shared understanding becomes the platform upon which their own questions are posed, and their own experiments devised. Questions like 'What am I interested in knowing? What might an explanation look like?' and 'How might I test it?' are in the end culturally informed. Our imagination in these things is constrained by whatever we take to be true at any given point in time. Therefore, most scientists most of the time are trapped within the dominant paradigm, taking the current models and exploring their potential, finding data to support them, or asking new questions on the back of them. Kuhn refers to this as a mopping up operation.

If the model is thought of as a vehicle travelling through space, some enjoy the unfolding view while others busy themselves with maintenance. What very few do is try to alter the ship's course, that they may view the universe from a different perspective. A whole range of questions are never explored, simply because scientists trained within a particular paradigm would never dream of asking them. And, in tandem with this, because the paradigm forms the basis of all inquiry, it is not natural to seek to undermine it in normal circumstances. Consolidation, not falsification, is the scientific instinct.

According to Kuhn's description, not only is refutation for a theory not actively sought, but when contradictory data does come in, it need not lead immediately to the theory being overturned. Historically, science has always been capable of living with a certain degree of problematic data. Kuhn points out that Newton's formulations of planetary movement did not coincide exactly with the recordings made via telescope, but they were still sufficiently accurate to impress. Nobody said, 'They're not exactly right, theory refuted.' He also reminds us of the vital role politics play in new discoveries. When Roentgen accidentally happened upon the X-ray, the powerful figure of Lord Kelvin dismissed the finding as a hoax. This dismissal found support in part because if Roentgen was right, many established scientists would be forced to redo their experiments, as they hadn't allowed for potential X-ray effects. Accepting new theories has implications in terms of status, in terms of economics and in terms of psychological comfort.

So Kuhn suggests that only very rarely do genuine revolutions occur. No theory, he maintains, can claim to be entirely consistent with known observations (in so much as eventually any explanation runs into mystery) and as such, if we were to reject all theories on falsification criteria alone, we would have no theories at all. But occasionally revolutions do occur. Eventually the level of anomalous data building up becomes too great to ignore, and a theorist comes along who suggests a whole new way of looking at the problem. There is a heated battle, and if the theorist is successful, the new paradigm forces us to re-evaluate our previous knowledge: essentially, to see things from a new perspective. The battles according to Kuhn are typically protracted and bloody. Or, in the words of the physicist Max Planck:

> A new scientific discovery does not triumph by convincing its opponents and making them see the light, but rather because its opponents eventually die, and a new generation grows up that is familiar with it.

Classic examples of revolutions in scientific thinking are Copernicus' model of the solar system, Newton's mechanics, Darwin's model of Natural Selection, Einstein's work on Relativity, or the field of quantum mechanics. In each case a previous model limped along, although more and more data was appearing which it simply could not explain. Eventually a new theory explaining this data emerged, and the next generation of scientists, finding it a better fit and hence more compelling, adopted it.

There is much that feels convincing in this description of science. The story of science does appear to be one of great upheavals, and a good deal of work seems to focus primarily upon verification. Kuhn's theory highlights the powerful importance of psychological and political influences on progress in a supposedly objective field, and it reminds us of the pragmatic needs of scientists. In the end, we do not have science just for science's sake. A new theory must serve some purpose, be it a technological application, the raising of new and interesting questions, or simply meeting the recreational needs of the problem solving scientist. So it does seem natural that scientists are likely to be more interested in exploring a theory's potential than they are in finding a better one. Otherwise they are in danger of becoming like the car enthusiasts who spend more time replacing their cars than driving them. Kuhn's treatment is appealing because it seems to resonate strongly with the way the world is. It is, dare I say it, a common sense rendering.

So then, does this mean Popper was wrong? Well, only if you think there is a choice being forced upon us at this point: either science concentrates on falsification à la Popper, or it spends most of its time on verification and consolidation, à la Kuhn. The trouble with this sort of thinking is that it ignores the possibility science might do both. For, if we do away with the preconceived idea that the two theories exclude one another (a preconception driven in part by academic turf wars) we find some interesting things. Let's start with the demarcation problem. Popper says a sufficient definition of a scientific statement is its refutability. Kuhn in turn says that science proceeds by way of paradigm shifts, and these paradigm shifts occur when the build-up of anomalous data becomes too great to ignore. And of course, to say data is anomalous is to say it can not be accommodated by the existing theory. In other words, it refutes it. So, although Kuhn does not explicitly articulate this, on the matter of demarcation, there appears to be much common ground. Neither theory can run unless scientific statements lead us to clearly expect some phenomena, and not others. Furthermore, Popper and Kuhn acknowledge that in real life, refutation is difficult to establish, so again this needn't be seen as a big stumbling block between them.

The next question to be addressed, therefore, is how does science get its progressive characteristic? Kuhn accepts there is progress in science, and indeed it is just this progress he is trying to explain. So here both men are working on the same problem. Popper emphasises that once we refute a theory, we move on, and this in time provides progress. Kuhn says that only when this refutation has reached a crucial point, the political forces have worked their way through the system, and a better competing theory has emerged, do we progress. So Kuhn is emphasising the practical, in terms of how revolutions occur, and Popper is emphasising the philosophical, asking what is the key element which makes moving on possible. Again, there is no essential disagreement. Refutation, the mismatch of data and model, is, according to Kuhn, the way the argument is eventually resolved. And Popper himself made clear the difference between a theory being refuted and it being rejected. He pointed out that rejection requires the emergence of a better theory, and so again, the thinking seems to be complementary.

Kuhn did not consider the philosophical aspects of truthfulness the way Popper did, and so had nothing to say on the induction problem, but to the extent the two theorists dealt with the same problem, the match between theory and observation remains crucial to both. Kuhn's description of the way science goes about its business, and the inherent warnings regarding the ways practitioners can unwittingly stand in the way of progress, are to me both fascinating and important. But those who have interpreted this treatment as a refutation of Popper impress me less, for they appear to be holding Popper to account over statements he never made.

Having said all of this, we do need to take Popper to task on at least one point. For, against the backdrop of what is essentially a pragmatic theory regarding the conditions necessary for scientific advancement, Popper threw a few splashes of idealism. He hoped not just to define science and explain its progress, but he also wished to answer Hume. I suspect it was this very idealism that attracted those who like their science with a healthy dash of religious zeal to his theories in the first place. So it is of some importance to note that in this endeavour Popper was less successful. Remember his solution to the induction problem: although we can never know when a theory is right, we can still sometimes be sure that it is wrong. A contradiction is a contradiction is a contradiction. Only it isn't.

Let's go back to those black horses. We see a single brown horse and say, okay, 'all horses are black' is now falsified. 'All horses are either black or brown' is therefore a better model, a closer approximation of the truth. But what if the brown horse you saw was the only brown horse in the universe, and it just died? Is it still reasonable to say that 'all horses are either brown or black' is a better description of the world than the falsified 'all horses are black' counterpart? Or consider the sequence 1, 3, 7. Is the next number 13 (add two, add four, add six) or 15 (double and add one)? Let's say we look at the next number in the sequence and find it is 15. So the 'adding evens' hypothesis has been refuted. But what if the fourth term is a rogue element, and after that all the subsequent terms revert to the 'add the next even' pattern? The falsified theory is a closer approximation to the truth than the non-falsified one. What's going on here? Well, Popper was assuming that the world wouldn't play these rogue tricks. In other words, he only got around the induction problem by assuming the world was indeed consistent. In other words he used induction. And that's cheating.

Does that matter? Well again, that rather depends upon what you think science is for. So this is a good time to pull in to the side of the road and check our maps.

The thing that Popper undoubtedly achieved, and for which he remains rightly famous, was to provide a way of thinking about the difference between scientific and non-scientific endeavour, a distinction which it is crucial to keep in mind as we progress into the second half of this book. Some descriptions of the world are falsifiable, and some are not. Yes, at the boundaries, distinguishing one from the other is difficult, but at the conceptual level, the difference remains clear. Some descriptions of the world, those which truly deserve the title scientific, make predictions about the relationships we are likely to observe, and some do not. The accuracy of these predictions is the mark against which science is judged. If we are looking for a way of choosing between the models offered by science, then this is surely it: how well do the predictions match the measurements we make in the real world? The model which provides the best match is the one we turn to, at least until a better option comes along. And the ability to thus refute a theory explains well science's ability to make such startling progress.

Once we have prediction at the heart of the scientific method, there is an important question to be asked of any model. 'What are the unique predictions made by this model, against which it can be tested?' From this, Popper, like Bacon, provides a warning about the nature of bad science. Bad science will be extravagant in its claims and unclear in its predictions. Bad science will build the model around the existing data, while providing no hints as to where new data might be found. Bad science will shape shift, adjusting constantly to fit the data, in the manner of an undergraduate Marxist narrative. Bad science will support two opposing camps, endlessly relitigating the same old issues, going nowhere. And above all bad science will be mystical, wrapping its everyday models in the most wondrous stories, claiming for its endeavour a sheen of poetic respectability. The second half of this book is about this fine distinction between good and bad science, and what happens when practitioners on either side ignore the science/storytelling divide.

Before we move on to that, we should try to learn a little something from Popper's failure. What he did not do was answer Hume's challenge: and as such, he doesn't give us good cause to think of science as 'approaching truth' in any useful sense. This in fact takes us to the second category of error I want to hunt down in this book. For many scientists are still Platonists at heart. Is the universe really governed by simple, consistent and explicable laws, or should we take a more Kantian stance, and think the design of our brains simply makes it convenient to see it this way? Our answer essentially becomes a matter of faith. We can be confident that scientific models provide good means of prediction, but not necessarily of description.

This breed of scepticism will, as ever, make some readers uncomfortable. Those who wish to believe that data can of itself yield meaning, those who rail against science being reduced simply to technology and storytelling, will cling to the possibility of greater meaning. Perhaps Steven Weinberg speaks for many such thinkers when he writes:

> The effort to understand the universe is one of the very few things that lifts human life a little above the level of farce, and gives it some of the grace of tragedy.

I can't help but wonder what these other 'very few things' might be. How about falling in love, reciting a poem, dancing, listening to music, getting drunk, flirting, watching a play, climbing a mountain, supporting a friend in need, attending the theatre, reading a novel, celebrating a World Cup final, delivering a eulogy or driving along a windy road with the top down…?

I don't want to suppress the romantic spirit with which many approach the practice of science. Far from it. To find meaning and greater purpose in our endeavours is surely the most satisfying achievement. The scientist driven by the depth of mystery and the nobility of the struggle for enlightenment is a fine role model. All I would beg is that we understand this *is* an inherently romantic stance, and not pretend these daring leaps of faith are supported by the same careful edifices of reason upon which the predictive part of the model is built.

This plea is going to make a lot more sense within the context of some examples from contemporary science. I've been bleating on about the importance of the science/storytelling divide for long enough. Now I need to get my cards out on the table and show exactly what I mean.

two

Part Two:

Heading Home

4 Of Cabbages and Warthogs

I was brought up in the Roman Catholic tradition, and remember with unfashionable fondness a small-town childhood of kindly nuns, certain priests, and strange rituals that always seemed to outlast my attention span. And I remember the stories. Everywhere there were stories. Great floods, smiting jaw bones, tempting apples, drowning soldiers, walking on water and in some mysterious sense, a man who died, but not really. It's impossible now, with the benefit of subsequent experience, to know what my childish self made of these tales. Probably at some stage I took them to be literal truths, in the way one might believe in Santa or advertising. And at some later stage I came to see them instead as stories.

One wise old head, during this transition from consumer to doubter, suggested to me this was exactly as the church intended. His argument was that the church's stories were simply devices used to introduce children to the realm of the ethical and the spiritual. No one believes these things actually happened, he contended; they are not meant as models of our physical existence. Rather, they are ways of ensuring children are brought up within a coherent moral and metaphysical tradition. Religion, by this view, has nothing to say about the way the physical world works. Instead it provides us with an opportunity to build up a satisfying sense of who we are, of our meaning and our obligations. By his take, religion stands or falls upon its usefulness in securing for a population a fulfilling spiritual existence and a harmonious social one. This in itself is an appealing story, and were I in the business of selling religion, I might make use of it.

To a teenager, however, with a questioning nature and a certain delight in combat, the explanation seemed to overlook two important points. Within the community of adult believers, there was no evidence the stories were viewed this way. Instead, most appeared to have succumbed to some sort of literalism, and those who hadn't apparently considered it polite to keep their feelings to themselves. Catholicism has no coming of age ritual where the old enough are whisked aside and told, 'Yes, it's all a convenient fabrication, just like Santa, but don't tell the youngsters, not yet.' Instead the church community, not only in my small town but across the globe, openly spoke of prayers being answered, physical resurrection, and real places called heaven and hell. Even today, new saints-to-be must await their miracle, despite the obvious problem that just as there are too many chimneys to visit, there are far too few miracles to go around.

The second flaw was perhaps more damning. If the aim of the institution was to establish a strong ethical base, why did it appear to a simple-minded western teen of the eighties that the ethics were so weirdly skewed in favour of the powerful? What exactly did the church have against women, homosexuals, and those of other faiths? We asked many questions of our teachers, and as I remember it the answers weren't all that compelling. Most of us drifted away, disappointed by a glimpse of what religion might have been, but clearly was not.

And yet, perhaps as with all of those raised in such a certain and untenable tradition, I am left with the legacy of insatiable curiosity. Human societies can't function without a sense of meaning, and rising from that, a sense of ethical obligation, from whence, in turn, we derive our shared sense of rules and responsibilities. What limits should be placed on ownership rights? What types of dishonesty are acceptable? Should we be able to use physical force in raising our children? Where is the line between personal ethics and the greater good? What is our level of responsibility towards those in other countries, or subsequent generations? To what extent should subcultures within a society be able to set their own rules and standards? What rights should we extend to other animals? We can imagine an almost endless list of this type of question, and science alone, devoid of a cultural context, doesn't appear to be able to provide the answers. So our storytelling traditions are left with a vital role.

But what happens when different story traditions clash, or our stories seek to deny the advances of scientific models? Maybe the very best example of this is the great evolutionary debate. In this chapter I want to look at the issue in more detail, hoping to show that the battle is as much a clash of competing and inflexible stories as it is a clash between scientific models.

The place where the clash is most apparent is within education. While many of us are happy to leave other people alone to believe any old thing that works for them, this sensibly relaxed approach proves problematic when it comes to schooling. State funded schools must, for reasons of efficiency as much as anything, settle upon a set of shared stories to tell. We need to somehow agree not just on what we want our own children to be taught, but by definition, what other people's children should be taught as well. With an issue like evolution, where there are two powerfully opposed viewpoints within the adult community, this becomes problematic.

The dominant scientific position in this debate is a fairly simple one. Science education, it is said, is about teaching models, not interpretations. And the scientific tradition has produced a method whereby it is possible, at any given time, to describe the best of the current models. These models will always have areas of uncertainty about the edges, but whether we're considering the structure of the atom, a model of the solar system, or the flow of genes through space and time, we will be able to give the next generation the best available explanation. This, the boosters claim, is the purpose of science education: to share with the next generation the most up to date descriptions of the world, and the methods used to reach them. And, the scientists add, evolutionary theory has thus far passed the tests. To suggest therefore that a markedly inferior model be taught alongside evolution amounts to intellectual treason.

Creationists (I use this blanket term for convenience) make a number of cases against this. The first is that evolution is not well established by scientific criteria, and so alongside its claims should be taught its weaknesses. The second, more fundamentalist approach is that the whole damned thing is a lie, and the world was created as per biblical explanations. Any attempt to challenge the word of God is nothing more than arrogance. Finally, and most subtly, there is the argument that there is an important difference between evolutionary theory, and the cult of evolutionism, and the latter must be exposed.

The thing that makes this argument interesting is that both sides have some useful things to say. Evolutionary theory is well established by scientific standards. There is no satanic plot to keep the truth from the people. Scientists are genuinely piecing together the evidence as best they can, and our current take on evolutionary theory is the picture which best fits the jigsaw pieces. But equally, evolutionism is a genuine phenomenon. Many of evolution's strongest public proponents are still caught in Plato's thrall, something that was very clear back at the beginning of the debate in the Victorian era.

To get our heads around these issues, and one proposed method of resolving them, there are three things we need to work through. First, an outline of the basics of evolutionary theory in all its glory. (This will of course be the most simplistic of treatments. For those of you unfamiliar with the ideas I introduce, check out the bibliography for writers who do this in far greater detail, and with a good deal more authority.) Second, I'll try to examine the scientific case for creationism, which I should say from the outset I believe to be neither scientific nor a case. And from there, if I am right and the case is fatally flimsy, it is natural to wonder how the controversy ever grew so strong. When an argument refuses to die, we should ask whether it is just that the details remain unresolvable, or if instead the surface debate masks a deeper and unnamed disagreement. I think the second explanation is the right one. To defend this view, I present for the prosecution the life and times of one Thomas Huxley (1825–1895), evolution's spin doctor, who from the outset ensured that the evolutionary battle would be as much tribal as scientific.

So, evolution eh? How to give a taste of its wonder in the space of a few thousand words? First, it is wrong to think of evolution as a single theory. It is an umbrella term, describing all sorts of connected hypotheses. There are theories on speciation, on the precise details of inter-relatedness, the nature of mutation, the distribution of variation through populations, natural selection, development, the first spark of life, the drivers of change, the pace of change, animal behaviour, the folding of proteins, interpretation of the fossil record, the nature of feedback mechanisms, game theory in nature … you get the idea.

As with any fecund branch of science, the scope increases with each passing day, and at each frontier there is excitement, discovery and argument. It is safe to say the way we think about evolution in fifty years time will be different from the way we regard it today; our current stance will appear naïve, and in many places misguided. So, to be for or against evolution in principle is sort of meaningless, a little like being for or against chemistry.

Nevertheless, such a debate of principles does indeed rage, and it must be about something. As far as I can tell, it is mostly about a reduced form of evolutionary theory: the thinking behind Darwin's first bombshell, which with Huxley's stirring came to challenge the Victorian orthodoxy.

Let's start with that orthodoxy. Look around the natural world and what do you see? A thing of wonder; an intricate, complex network of self-sustaining life forms. Every living thing apparently has its place; its detailed design allowing it to fly, scurry, eat, reproduce, and grow old as it moves through an environment for which it is so splendidly well suited. So how on earth could that all come about? The old explanation was the one our instincts still strain towards. This sort of thing couldn't just have happened, right? It must have been made by something, or someone. 'When I see a watch, I think of a watchmaker,' goes Paley's old argument. And so too, life's splendour, diversity and design led us towards an obvious conclusion. Someone or something made this all happen. And what's more, the design was careful and deliberate. The world we see is the world as it was created: by Aristotle's prime mover, with a divine purpose in mind. The ubiquitous creation myth reminds us what a very natural leap this is.

But as early as the eighteenth century, cracks began to appear in the literalist creation consensus (or you could go back well over two thousand years, to the Greek philosopher Anaximander, who was of the opinion that we had all evolved from fish). Have a look at the following passages, all written in the century before Darwin's big idea.

> Many worlds might have been botched and bungled, throughout an eternity, ere this system was struck out, much labour lost; many fruitless trials made, and a slow but continual improvement carried on during infinite ages in the art of world making.

This was Hume again, here arguing that blind chance is indeed a serious rival to a careful, crafting God, when it comes to explaining creation. All you need is enough time. A couple of French thinkers were even more specific.

> One could equally well say that the ape is of the family of man, that he is a degenerate man, that man and ape have a common origin; that in fact, all the families among plants as well as animals, have come from a single animal.

This being Buffon, in 1753. And Maupertuis got perhaps closest of all, two years earlier:

> They would have owed their origin merely to some accidental formation in which the basic parts would not have preserved the arrangement present in their parents. Each degree of error would have created a new species, and from these repeated variations would have arisen the infinite diversity of animals that we see at the present day.

It is an interesting historical quirk that Maupertuis was discouraged in his musings by the satirical viciousness of Voltaire's retorts. Voltaire believed in fixed forms, and was sufficiently powerful and brutal to thus forestall the inevitable. Beware the articulate bully.

Nevertheless, Voltaire's victory was only to be a temporary one; human curiosity was leading inexorably towards the evolutionary conclusion. Why at this time? What was driving it? Well, this was the age of enlightenment. On the back of Bacon's certainty that nature could be forced to give up its secrets, close observation was the order of the day. And close observation threw up some tantalising riddles.

One of the first patterns to so titillate was the similarity in design between creatures. As long as we've had words, we have dealt with the world by categorising it. A child can understand that a fish is different from a bird, which is different from a reptile, an insect or a mammal. Life does appear to come in types. And within these broad categories more fine grained distinctions also seem quite possible. A horse appears more closely related to a donkey than it is to a mouse, and a human has more obvious commonalities with a chimpanzee than with a tiger. Tigers seem close to lions, and a mountain daisy appears to be a long way from the mighty kauri. However, as thinkers turned their minds to formalising these folk intuitions they found the process deceptively tricky. For the more detailed the distinctions become, the more difficult species are to categorise.

A dolphin at first glance has much in common with a shark, but peer below the surface appearance and such confidence quickly dissolves. A bat is like a bird in that it flies, but far more like a mouse in many other respects. The boa constrictor has, within its skeleton, the rudiments of a pelvis and hind legs, so what does that make it like? The dolphins' flippers share their bone structure with the human hand and the bat's wing. The ear bones of a mammal are a variation on the theme used in the jaw bone of a reptile. The sorts of solid boundaries upon which authoritative categorisations could be made are devilishly hard to find. One response to this was to say, 'Oh well, that's just how it is. Categories are useful at the broad level, but meaningless beneath.' But that is not the way of the scientist. The scientist instead thinks, 'Perhaps the very things which are making this task so difficult are clues to an underlying order which is not yet apparent to us.' As we've already seen, this refusal to bow down before the inexplicable, our determination to keep chipping away, has yielded tremendously powerful results.

And chip away we did. By the nineteenth century the notion of design by master plan was gaining popularity. Once the master plan for fish, say, is established, then all the fish created can be thought of as variations upon this plan. Or so the theory went. But the thing about science is, as you begin to answer one question, it begins to throw up others. Why, people couldn't help but asking, has the creator, in all his wisdom, chosen to do it this way?

In order to get to the bottom of these master plans, scientists undertook meticulous dissections, aided by microscope technology. The more that was revealed, the more mysterious it did all seem. Why are some options never explored in nature? And why are other options, not obviously optimal, apparently compulsory? Where are the animals with wheels? Or those that communicate telepathically? Why would a creator design such a useless spine for an upright creature, one incapable of withstanding the leveraged forces it is subjected to? And why for that matter have air and liquid taken in via the same tube, ensuring that on a regular basis members of our species choke to death? Why can't people fly like birds? (How cool would that be?) Why are humans so spectacularly helpless at birth? Why do the same solutions, many of them apparently ad hoc, keep showing up across such a diverse range of creatures? Why has creation been so stubbornly unimaginative? And what is a species exactly? If for any life form it is often possible to detect another slight variation on a theme, at what point precisely does one species begin and another end? What is the elusive essence of cat, that makes identifying one so simple, even when it is blind, lame and does not meow?

New and exciting discoveries added to the evidence and deepened the conundrums. Global exploration meant hitherto unknown creatures from far off lands were examined in great detail. And patterns of species distribution in turn cried out for explanations. Geology provided evidence that the earth was far older than anyone had imagined (but nowhere near as old as we now believe it to be). And this ancient land was beginning to yield up fossils of creatures from earlier times, creatures both strange yet weirdly familiar. More and more pressure was being applied to the literal creation story, which nudged science in search of an alternative explanation: one that could make something of the massively complex yet oddly regulated pattern of life.

The line of reasoning now seems obvious, but at the time it took a feat of wondrous imagination: if the earth was indeed very old, and if in the past very different creatures roamed the globe, and if, as we peer further back in the fossil record, less complex life forms appear to dominate the record, and if many living creatures have not only close relations living near them, but also exist with slight variations on the theme in the fossil record, and if many of the variations reflect geographical proximity, and if at heart we all seem to be deviations from the same theme, well couldn't we all be … related?

And so we stumbled towards the first great thread of the evolutionary story. All life on earth, it was postulated, was related, and furthermore these relationships were nested. All this means is that once you've distinguished animals, say, from plants, then within the animals you can distinguish birds from mammals, and within mammals you can distinguish marsupials from the rest and so on. From this it is a brave but not unimaginable leap to a second startling possibility, that this relationship reflects common descent. Maybe all life had a common source, and over time life forms had split and developed separately, again and again, until the diverse, but nested, relationship of the present day.

But this idea still left some fairly big questions unanswered. In particular, why should the change occur over time in the way we see, and even more important, how does such change occur at all? The first question gets to the heart of gradualism, and it is one of those obvious points that is easily overlooked. You could have a process of common descent, which is to say branching change over time, without actually retaining any signs of this process. For this to happen all that would be required would be a preponderance of large scale overhauls, where suddenly a whole new design is happened upon. Think of a car manufacturer introducing a new model. If the change is slight (many of the features of the old model are retained), then we have no problem tracing the lineage. But if by contrast the change is a total overhaul, for instance in the replacement of the iconic VW Beetle by the VW Golf, we have to rely upon the badge and our contemporary knowledge of the company involved to pick up the models. The very fact that we can read the relationships between various life forms in their design plans suggests that evolution has in general made incremental improvements (or been Porsche 911-like in its workings), and that leaps (the Beetle to Golf version), starting again from scratch, are either rare or non-existent. But why? Why hasn't the creation process gone for the big leaps? What is constraining it?

The answer to that question can only come if we understand the mechanism of change (or alternatively, from a 'full stop God' solution: 'It's just how He did it okay, so stop annoying me and go and play with your friends'). But as we have seen, thinkers like Maupertuis were already on the way to providing an alternative explanation. And, with Voltaire long dead, and new evidence accumulating, the time was right for Natural Selection. The fact that two thinkers, Alfred Wallace and Charles Darwin, happened upon the solution independently, only reinforces the view that this too was a process of gradualism. The key ideas were already known, it was just a matter of a suitable genius putting them all together. Darwin is regarded by history as the more important figurehead than Wallace, his explanatory framework and wealth of experimental data providing a more satisfying whole. His social status, moderate politics and influential contemporary supporters no doubt also played a part. (And of course, history, like evolution, has its self-reinforcing mechanisms. Once a hero has been chosen, it becomes easier to uncover the details of their legend, and so add to it.)

Interestingly enough, natural selection took a long time to be accepted, and was subject to endless misunderstanding, a situation which today is only marginally improved upon. So, indulge me while I linger a little on this wonderfully simple and profound mechanism. Remember the question we are trying to answer. How is change achieved down through the ages, from an initial extremely basic replicator, to the kaleidoscope of design we observe today? What, or who, is driving the process? The disturbing answer, as provided by the mechanism of natural selection, is nothing at all. Or rather, nothing but what we might think of as blind chance. The development of complex life forms, according to Wallace–Darwin reasoning, is no more miraculous than the flow of a river or the turning of a tide. It is the simple result of a basic law of nature. (Although one may still find miracles in the existence of these laws, if it's miracles you're after.)

So, how is it meant to work? Darwin began his ponderings with two vital pieces of information already in the bank. First, offspring tend to inherit some of the characteristics of their parents. Animal breeders knew this, and made much of it in practice, and from time immemorial people have observed physical qualities both good and bad being passed on down through families. Darwin had also read Malthus, who had written on the limit placed on population growth by resource constraints. Members essentially compete for their share of resources, and as populations rise and resources remain stable, many are doomed to fail in this struggle. Darwin's big idea, perhaps the most far reaching ever conceived of, was to bring these two elements together. In hindsight it may seem an obvious step, but no one else (apart from poor Wallace) had thought of it.

Let me try to explain the mechanism by zeroing in on a particular life form: the humble rat, familiar to all and detested by many. The rat's greatest crime in the end is only its remarkable success. It is a born survivor, seemingly purpose built for the modern world: physically robust, omnivorous, a ferocious breeder with a cast iron constitution and more than happy to make its living off our plentiful waste. If you were to design an urban survival machine of a certain scale, you could do a lot worse than using the rat as a template. (After a nuclear holocaust, they liked to tell us in the early eighties with cold war hysteria spluttering to a close, only the rats and the cockroaches would survive. A truly inaccurate forecast, but the metaphorical point was clear: anything that leaves such nasties to inherit the world is clearly evil — as if an explosion that spared all but budgies and kittens would be somehow kinder.)

How is it the rat got to be so very good at what it does? Natural selection requires that within a population there is some mechanism for introducing random change, or what we now think of as mutations. So amongst a given population of rats, from time to time one may be born blind, or with shrivelled hind legs, or a defective immune system, or particularly keen hearing. And indeed we know such things happen, so there is no controversy here. Now, in the first three cases, the outlook for said rat is not particularly good. Unable to see where it is going, unable to get there if it can, or likely to be struck down by disease upon arrival, such rats quickly perish, and the mutation is lost to the system. But how about the one with sharper hearing? It perhaps has a slight advantage in detecting predators, and so its chances of living long enough to pass on the mutation to its offspring are ever so slightly above average.

Now, if the change is heritable, then some of the offspring will also possess the same mutation, have sharper hearing, and so on average will be more likely to last long enough to slightly outproduce their competitors. In a world of limited food, aggressive predation, competition for mates and yes, even nasty mind-altering little parasites, every advantage helps, no matter how slight. Over time, as a result of the power of exponential growth, that slight advantage is all it takes. The mutation spreads through the population. Average rat hearing improves.

The system is not foolproof nor particularly efficient — a history of extinctions attests to its wastefulness — but it does provide an explanation for unguided change over time, and the accumulation of complex design.

Note too that the system is free of causal direction. Changes in design do not occur to enhance reproductive success. Rather the reverse is true: because some changes do lead to reproductive success, they are by definition incorporated into the design of subsequent generations. The process is as blind as it is powerful, producing observable patterns in life in much the same way as wind produces intricate patterns on the surface of a sand dune. It is a spectacular manifestation of the laws of physics, nothing else.

If you are like most people who encounter this algorithm for the first time, you may well have baulked at the suggestion that out of this we can get not just change, but also complex design. Really? Isn't there a world of difference between refining a rat's hearing and making the leap from amoeba to warthog? The difference, from the evolutionary perspective however, is only one of scale. Sharper hearing reflects a single change. Amoeba to warthog represents a long series of changes. All that is needed to turn one from the other is time, the three billion years or so since simple life first appeared on the planet.

Three billion years. Note how easily your eyes skipped over this number without taking the time to truly appreciate its scope. There are about as many seconds in a human life time, and a long one to boot. Think of that. Every second of your life represents an entire year. In a year a bacteria like *E. coli* can go through 26 000 generations. Now imagine that happening for every second of your life! That's an awful lot of trial and error, an awful lot of experimenting. At that scale, you almost want to look at natural selection and say, 'Is that all you've managed to come up with?' But, thinking on this sort of scale is not natural to us and I still have trouble with it, the same sort of trouble I have getting my head around the wonders of modern computers, built as they are upon the simplest of frameworks, a network of on/off switches. It seems impossible to believe such a simple system can underpin the playing of music and video games, editing films and surfing the internet, until one considers the complexity of the networks, the millions of on/off states that can be altered within the blink of an eye. So too with evolution. Only with scale does the whole thing begin to sound plausible.

Plausible is one thing, scientifically respectable is quite another. In science we don't just accept a theory because it sounds cool. Over time the scientific tradition settles upon criteria that work well when producing predictive models. We ask that any new idea is consistent with other known theories. We ask that it has explanatory power, that it provides a framework for thinking about new data. We ask that it makes testable predictions, so that we can reach objective conclusions about its power, as Popper suggested. We ask also that it is as elegant as possible, both because simplicity is easier to work with, and because history shows the power of reducing phenomena to simpler levels of causation. And finally we ask that a theory provides fertile ground for new research and speculation, which is to say, that it is in some sense useful. Crucially, we ask that it does these things better than any competing theory. When people claim that evolution is good science, they mean that they think it does well against this sort of check list. Are they right?

In terms of consistency, core evolution (common descent, design by natural selection) is an excellent theory. This in fact is its great appeal. It explains life on earth only by invoking the known laws of physics and chemistry, and the simple (though not in a philosophical sense) notion of randomness. It is able to explain the details of creation without turning to any new forces outside the realms of current science. The strength of this is that we are not in the position of having to say, 'One story is as good as another.' This is the only story that works with well established principles, and doesn't rely upon an imaginative leap into the unknown.

Evolution scores equally well on explanatory power. The very puzzles which led to the question are answered by natural selection. Because selection can only work off existing plans, and because each step forward must represent an improvement, we can begin to see why the same bone structure might be found in the wing of a bat or the flipper of a dolphin. We can explain the puzzling weaknesses in the design of the human body, and solve the riddle of the snake's legs. Behaviours that may appear strange — the sacrificial attacks of bees protecting their hive for instance — make sense within the mathematics of population genetics.

Despite a history of challenges in the area, evolution in fact scores very well on predictions. Evolutionary theory predicted that we would find a unit of heredity, and in due course genetic inheritance was discovered, and later the structure of DNA itself was modelled. Evolutionary theory predicted that at the molecular level, patterns in DNA would reflect known lineages, and so far so good on that front. It made predictions about finds we would expect in the fossil record, and again, no contradictory evidence is yet to hand. It makes predictions about how diseases might mutate, and what we might do to thwart them (look at how the world holds its fear-primed breath, waiting for the bird flu to mutate into a more dangerous form). It made predictions about the link between DNA and development, so spawning one of the most fertile areas of modern scientific research. It makes predictions about the sort of complexity that will emerge if we run computer simulations of undirected mutations subject to selection pressure. Again, such simulation results deepen the theory's plausibility.

Those who like to claim evolution is not predictive are often referring to the difficulty of using natural selection theory to tightly predict outcomes in complex natural environments. But this is a bogus charge. After all, a teenager is at heart prey to the laws of physics. That these laws can not in themselves predict how the said teenager will behave in any given circumstance hardly challenges the validity of physics.

Elegance is probably where evolution most shines. Natural selection is divinely simple. So simple indeed that I was able to outline it in a couple of pages, using nothing more than a reference to rats. Yet carefully applied, the selection mechanism provides an explanation for the complex design we see all around us.

The fecundity of evolutionary theory can hardly be doubted. From medicine to biotechnology, game theory or computer programming, evolution today seems to be everywhere. Darwin's big idea, by allowing us to shift our perspective from a static, designed world, to a dynamic, emergent one, did what all great science does. Ultimately it allowed us to build new models, make new predictions, and change the way we interact with it. It's given us plenty to think about along the way.

At first glance, evolution seems to pass the 'Is it science?' test with flying colours. This conclusion is strengthened when we consider that the thing we most ask of a scientific model is that it does its job better than any of the alternatives. And the truth of the matter is that there is no other explanation queuing up to do the work, or at least none that could meet the criteria above. If we are careful to confine our definition of science to predictive knowledge, then we can say that evolution is a better driver of biological discovery than any other model of life on earth. And that really is saying something.

Nevertheless, unlike, say, our models of chemical reactions, evolution is dogged by public controversy. Predictably, I'm most interested in whether or not this controversy has something to do with storytelling.

It is worth noting that evolutionary controversies naturally fall into two distinct groups. On the one hand, there is the range of disputes taking place within the scientific community. For scientists, while all agreeing on the mechanics of evolutionary theory, appear happiest when disagreeing on what it all means. For the most part these arguments are best viewed as the working through of the Popper/Kuhn mechanisms of progress: wrestling with metaphors, challenging methodologies, proposing new models and defending old ones. These debates are often technical, and while fascinating in their own right, this is not the place for them. (I'll make an exception for the debate about the relevance of evolutionary theory to our understanding of human psychology in the next chapter: because it's sort of fun, and a background in education leaves me feeling as if I might have something useful to say.)

For now I want to concentrate on the other hand: the more headline grabbing of the disputes, where an aggressive and well-organised group of doubters are attempting to cast doubt upon the mechanics of evolution itself. I am going to make no effort at all to pretend to be impartial here. Try as I might, I can't generate any real sympathy for the arguments of the Creationist lobby. Their arguments seem plain loopy to me, and I'm going to try to explain why. Then I want to explore why the debate ever got to be so fierce, given the flimsiness of the basic arguments. You will not be surprised to find me pinning the blame on a serious misunderstanding of the science and storytelling divide. You may be more surprised to hear that it is a misunderstanding of which both sides are equally guilty. But one thing at a time.

Many millions of people around the world (disproportionately represented by the highly educated population of the US) are convinced that all I have told you about evolutionary theory thus far is entirely incorrect. They don't just think the model is sometimes prone to misinterpretation. They think the whole thing is plain wrong. By their thinking, either I have been misled in my research, or I am deliberately misleading you. I want to take you through what seem to me to be their favourite arguments, just because if I didn't, you might wonder what it was I was frightened of. (Some approaches, like spurious claims to support from the laws of thermodynamics, I have deliberately ignored. Life is just too short.)

One of the oldest objections, which has shown surprising staying power, is that of the 'gaps in the fossil record'. The fossil record, the argument goes, is just far too incomplete to show the sort of progression evolution requires. Where we might expect to find intermediate forms, we find only gaps. According to this argument, there have been many weird and wonderful creatures on the planet over time, but they have not morphed one into the other. Rather than gradual stepwise change through time, the doubters envisage a series of creation events, with each life form able to vary only within a narrow range of characteristics. They claim that the fossil record, patchy as it is, supports this view.

At this point it is customary to veer off into long and tiresome discussions on particular aspects of the fossil record.

'That's not a horse.'

'Oh yes it is.'

'So why's it got so many toes then?'

'The modern horse has lost some.'

'How?'

'They've been selected out of the population.'

'No they haven't. It became extinct, and then God made a new one, with fewer toes.'

'He did not.'

'Prove it.'

I think there might be a short cut available here, which will avoid this dance of detail. You see, I think the 'gaps in the fossil record' argument has a logical flaw in it, and it makes the other technicalities irrelevant. Let me explain, with the help of an analogy.

You are wandering about a strange town, and notice that between two houses there is a mound of ash and twisted debris, the result, you assume, of a fire. Because the ex-building sits between two houses in a residential block, you assume it too was once a family home. You check with a neighbour. They have only moved in since the fire, but tell you they heard it was a family home which burned. You consult with the city plan, which shows that a house was built on this spot thirty years earlier. It seems reasonable to assume a house burnt down, but you can't be sure. Maybe it was a factory, or a commercial property of some sort. You are strangely curious about this, and so sift through the wreckage. You find a few artefacts. A family photo perhaps, an old bath tub, and miraculously, the head of a doll. You're now fairly satisfied. Family home is the best explanation.

'Wait a minute, though,' says your doubtful friend. 'You can't possibly reach that conclusion. What about all the gaps in the record?' 'What do you mean?' you reply, a little puzzled. 'Well,' they patiently explain, 'family homes have beds in them, televisions, couches. You have found no evidence of any of these things, so you can't assume it was a family home at all. You record is far too gappy.'

Do you see the problem? Time, like fire, destroys evidence. That's sort of the point. The fossil record represents the tiny proportion of life that leaves its mark, the equivalent of the stray photograph, the preserved bath. The gaps are not excellent evidence of anything, other than fire. Should the sifting produce conflicting evidence — a bank vault or an industrial printing press, say, or in the case of the fossil record, a bunny rabbit skeleton in the Cambrian — then of course we'd be forced to rethink our position. But, with no such evidence emerging, and ample supporting evidence from other sources (for the fossil record has never been the major support of evolutionary theory, just one of the bits) we are bound by the laws of reason to stick to our hypothesis.

So far all the fossil evidence we have found supports the evolutionary hypothesis. There is no contradictory evidence, despite the fact that we can confidently describe what such evidence could look like. There must, by definition, be gaps in the fossil record, irrespective of whether or not the evolutionary explanation is the best one. The 'gappy fossil record' argument is a nonsense.

The second argument suffers from a different but no less fatal fallacy. This is the argument that while natural selection pressures can refine and alter the characteristics of a population, this alteration can only occur within species' boundaries. That is, evolution may make a rat more robust, but there is no mechanism by which it can turn it into a cat. This is an odd sort of argument, based as it is around a concession which, when followed through to its logical conclusion, collapses the whole case. The concession is forced upon the defenders by the rather obvious fact that the selective breeding of dogs for instance, has produced a range diverse enough to encompass both the Chihuahua and the St Bernard. Nobody, as far as I know, argues that different dog breeds haven't been created by, well, breeding. And there is no good reason for supposing that a process guided by human intervention could not also be achieved, albeit over a larger time scale, by natural selection.

However, the Creationists make the Aristotelian claim that this process faces a natural barrier. A dog, they say, can be reshaped in many ways, but it will still be a dog. There is an essence of dogginess that no amount of mutation and shuffling will remove. In order to make anything sensible of this argument, we must be able to work out what exactly they mean by this essence. What would have to be removed from a dog before we agreed the essence had gone? Its wagging tail? How about the bark, or the tendency to chase cars, crap on lawns or sniff other dogs? The simple truth is that no such essence exists. A dog is a dog because that is what we choose to call it. It is a convenient category, constructed by us, for our own purposes.

One way to define species in biology, is to think of them as being reproductively isolated, the reason being that at this point there is no flow of genetic material between the populations. If we take this as our barrier, then the news for the Creationist is all bad. Quite apart from the fact that St Bernards and Chihuahua, left to their own devices, are not in the habit of interbreeding, lab studies have also shown how separated populations of fruit flies can over time accumulate mutations which cause them to become reproductively isolated. So by this definition, one species can split into two. Ah, but they're still fruit flies, protest our Creationist friends. Okay, fair enough, they still appear physically similar, for now. But once reproductively isolated, what's to stop them accumulating physical changes in the manner of, say, a Chihuahua and a St Bernard, which the Creationists allow? With reproductive isolation and dramatic physical change both clearly demonstrated, where is the species barrier hiding?

In the wild we see examples of geographically-spread creatures, gulls in Europe for example, or moths in Asia, where each subgroup is able to interbreed with adjacent populations, but not with those from further afield. Species becomes a fluid definition, and the notion of essences dissolves into biological meaninglessness. The recent ability to create transgenic species underlines this point. The argument for species barriers is founded on a poorly-drawn definition. As soon as the definition is tightened and Creationists are forced to define their barriers, we are able to show examples of the barrier being successfully negotiated in nature, and the problem disappears.

Next we must turn to another ancient objection: that of irreducible complexity. The most recent proponent of this view is Michael Behe, who makes his argument at the molecular level. Despite being trumpeted as a new attack on the evolutionary orthodoxy, it is in fact the redrawing of an argument that has been at the heart of objections to natural selection since Darwin's time. Natural selection remember, achieves complex design by altering the unit one mutation at a time, with each step being a more effective path to reproductive glory than the previous. Aha, say the objectors, but how can this possibly be? For what use is half a wing, or an incomplete eye? At the superficial level, both these challenges and others like them have been well met. Half a wing may not get you off the ground, but it can certainly help you glide from tree to tree, a trick used by mammals and reptiles alike. Or an insect's wing may evolve first to suit some other purpose: as a solar heating panel for instance. Similarly, a little sight is an excellent thing in a world where your competitors lack such an advantage. In the land of the blind, the one-eyed man is indeed king. And the various eye solutions nature has hit upon all involve designs that could be made through a stepwise refining process.

The Behe attack is different in degree, rather than in kind. He quite rightly observes that the truly wondrous complexity occurs at a far lower level of analysis, down amongst the molecules where just the right proteins have to be present in just the right combinations in order for the many micromachines that make up our body to work. And again, he wonders how on earth any particular mechanism could advance piece at a time, when it appears that each and every component is vital for its eventual functioning. It is a good thing to wonder, and represents part of the wide field of puzzles modern evolutionists are attempting to solve.

To date, some of the apparently irreducible mechanisms highlighted by Behe have indeed been reduced, and in fact, the solution had in some cases been found well before his book went to print. So irreducible is certainly a misnomer. The expectation is that this will continue to be the case, based upon the very scientific notion of 'so far so good' (remember: give up this principle and you have to give up all science). Up until this point, 'selective' thinking has proven a remarkably fertile approach, and again there is no serious competitor on the horizon to test natural selection against.

The logical problem with the irreducible complexity argument is there's no coherent way of distinguishing between 'this can not be broken down to a simpler version and still operate' and 'I personally can not think of a way of breaking this down to a simpler version and still have it operate'. Behe claims to be arguing the former case, when in fact he is proposing the second. It should not surprise us that proof of irreducible complexity is beyond him, for the logical path he has chosen is again flawed in its construction. There is no consistent way of arguing from the absence of apparent mechanisms through to the impossibility of such a mechanism existing. The proof of impossibility, upon which his whole case rests (and if it could be shown it would indeed blow evolution out of the water) can not be built this way.

All of science, if we dig deep enough, is based upon mechanisms we have not yet explained (gravity is an excellent example). And as the earlier chapters showed, science never relies upon perfect explanations, just the best so far. Behe's objections are a call for further research, not an excuse for abandoning the theory. In science we abandon a theory when a superior (more useful) alternative proves its worth. Yet, the Intelligent Design lobby have no testable theory waiting in the wings as a replacement. They're just a bit troubled by the unexplained parts, which leads one to wonder why they are not attacking physics or chemistry with the same sort of force. The answer of course is historical, and I'll come to it later.

Finally, I come to the 'but that's really, really unlikely' argument. One of the hardest parts of the evolutionary picture to fill in concerns the emergence of the very first replicating units — whatever they were made of — in a world of chemicals. At some stage, clusters of molecules with the property of self-replication must have emerged: the mystery is in the detail of how this happened. Given that we don't know exactly when, where, or how it took place, the Creationists see it as an excellent place to launch an attack. Hence, some argue that while the path from first replicator to biodiversity 2007 may have been evolutionary, the initial spark was designed (other people argue it came here from outer space!). Here the motivation should probably be questioned. If you're going to go back that far, why not cede some extra ground and plonk your intentional creator down at the emergence of the conditions of existence, where he's far less likely to be disturbed by scientific discoveries during our lifetime? Where exactly design-supporters choose to put their designer seems to carry some flexibility, without the meaning of the story being seriously challenged. Once again, storytelling can to a large extent float free from discovery and settle wherever its needs are best met. So, how good a place is the beginning of evolutionary time?

It's not a place I would choose, purely because the current scientific orthodoxy is getting so close to it that before long you'll be forced to move again. It's a little like having your sunbathing interrupted by a rising tide. You can move your towel back a few metres, but it's only a temporary solution. Making camp just above the high tide mark makes much more sense to me. We don't know exactly how life started all that time ago, but we do know enough about the prevailing chemical conditions to do some informed guessing. Plausible models of random generation have been proposed, but for now the information we have is insufficiently rich to do much more than say which guess appears the most likely. And into this uncertainty the Creationists wish to throw another possibility. What say another story, one of an intelligent being helping the process on its way, is a better one?

Scientists are reluctant to embrace this approach on the grounds of consistency, for to accept it would represent a sort of 'game over' point in their discoveries. Having clawed their way back to the very beginning of life, using nothing but the evolutionary mechanism to guide them, they are being asked to stop and accept that 'at this point, some magical stuff happened, courtesy of a big mystical force I've just made up because I like the sound of the story'. Knowing how far their methods have taken them, it is perfectly natural, I think, for the scientists to want to keep going.

The random generation of life as currently described is plausible, albeit not in itself highly likely. So, do we seek to learn more about the conditions at the dawn of life in order to refine our description and reduce the odds, or do we just give up and say, 'Enough already, we need to leave a place for God.'? Here we are getting into the realms of faith and philosophy, rather than science. Scientists by their nature wish to keep on pushing, and we the non-scientists must surely respect the fertility of this approach. Is there something wrong with an individual stopping at some point, and saying 'I don't want to know any more'? Of course there isn't, so long as they acknowledge this is what they're doing.

Which brings me in a round-about way to the likelihood argument. Some Creationists not only want to stop at this point, but they want everybody else to stop here too. They argue that no solution will ever be found to this problem, because they've done the figures, and the chances of life spontaneously assembling itself are just too small. I have a few problems with this approach. The first is, I'm not comfortable they have done the figures at all. I don't think we yet know enough about the initial mechanisms and conditions. (Did it happen in the clay at the edge of the water, or far beneath the earth's crust for instance? Both possibilities have their supporters.) And I'm also puzzled by the 'it's just too unlikely' argument on purely logical grounds, for it presupposes that there is some threshold of likelihood beyond which we must declare a process impossible. But where exactly is that threshold, and why do we choose to believe in it?

If I was to walk into a casino and throw twelve dice in a row, and they were all sixes, you would be astounded, and suspect I cheated. And so you should. The chance of it happening is only one in two billion. Yet, and here's the rub, if you were to pick up a dice now and roll it twelve times, recording the numbers that come up, you too will have just thrown a combination with odds against of one in two billion. The reason is simple, but perhaps not often enough dwelt upon. You have to throw something — blame gravity and Newtonian mechanics for that — and there are two billion possibilities. Of these, one simply must occur. Events which are declared extremely unlikely in hindsight, are not as impressive as our instincts tell us they should be. In a world of endless permutations, near-impossibilities are inevitable. I don't see twelve dice falling randomly to the ground and immediately assume God must be involved, so unlikely is the outcome. For in this case, an extremely unlikely outcome is inevitable: a nice little paradox we need to bear in mind.

Scientists themselves are often tempted by the same category of mistake. Cosmologists, when they realise that the conditions required to make the universe as we know it are so remarkably precise and unlikely, are disturbingly prone to go all weak at the knees and make reference to higher powers. But, is this also just a dice-like situation? Are we being unduly impressed after the event? My hunch is that we are, but it remains an interesting area for further philosophical inquiry.

In the meantime, I think it's reasonable to demand a certain consistency in our musing. If our argument is to hinge upon likelihood, then we must make sure we ask, 'Is the explanation being proposed more or less likely than any other option on the table?' Random life generation is unlikely, but at least we can discuss how the odds might be calculated. The creator version runs into problems when we try to apply the same sort of thinking. A transcendental power creating something out of nothing, by methods which we can not understand, when methods we do understand were available: how likely is that? Is there any way of knowing? Crucially, is there any sensible way of deciding when unlikely has become too unlikely? I know people who are spooked by even the mildest coincidence, certain that it must have some sort of deeper meaning. But as always, meaning is about personal preference, not evidence. Those who have resisted spooky explanations have discovered some fairly amazing things in the past, and when it comes to modelling the world, I'm all for going with the methods that have the strongest track record.

So, enough of the Creationists already. As you can see, I am not particularly moved by their arguments, and neither are the vast majority of biologists (no, well they wouldn't be would they?). And given my belief that in the end their arguments collapse under their own weight, I become interested in another question. Why is it that the debate has generated such remarkable fury, and for so long? Why hasn't it, like the Theory of Relativity, just been quietly accepted as one of those technical things sitting there in the background for those who are interested to play with it? Why are there no lobby groups with placards demanding that we point out the gaps in our understanding of gravity in the physics classroom, or that we warn students of the controversial nature of infinity when teaching mathematics? Why is evolution alone singled out for this curious treatment?

The answer I would give is a predictable one. To me it's got nothing to do with the science and everything to do with the story. We, as human beings, need our creation myth. It is an essential part of our make-up, some sort of a design quirk if you like, that relates the 'What does my life mean?' question to the 'How did I come to be here?' question. The link between them is not made necessary by logic, but rather by our nature. It is the most natural link in the world for us, as reflected by the enduring popularity of Aristotelian thinking. To understand what we are, we think we must first understand the purpose with which we were created. That means any model which speaks to the manner of this creation will cut to the very heart of who we think we are. It is extremely difficult to debate the evolutionary model because it is almost impossible to prise free the stories that have formed about it.

This is clearest in the religious context, where certain fundamentalists explicitly link God's means of creation with his intentions for the created. Less obvious though, and therefore potentially more insidious, is the way the scientific community often allows their own personal world views to piggy back into town on the evolutionary model. This common error has been termed evolutionism by some, and the need to expose it as fraudulent is just as pressing as the need to resist creationism's whacky claims.

Let's just take a look at the way evolutionism's case is constructed. The first step simply involves establishing evolution's credentials, as I did earlier in this chapter. Look, evolutionists say, here is a mechanism by which random mutation and a combination of natural selection and random drift can work together to produce the great variety and complexity of life we see about us. Therefore it is no longer necessary to evoke a higher power as the direct creator of this awesome order: a perfectly acceptable alternative explanation exists. And that, in so far as it goes, is perfectly reasonable.

But now observe carefully the switch from predictive modeller to storyteller which so often takes place. Suddenly the narrative changes, describing not just the mechanism of change, but also the implications of this mechanism. We are here as the result of nothing but random chance, we are told. The universe was not created with us in mind. We were not meant to be here, we simply happened, a sheer fluke, the result of contingency and the laws of physics. The universe has no attitude towards us, it is nothing but physical relationships. Creation is random and disinterested. There is no God, no purpose, no meaning. We simply are.

This is a natural leap to take at this point, of course. Because the old models of meaning were built about an erroneous model of creation, it is quite predictable that some should associate the replacement model with the opposite conclusion. Perhaps this is what George Bernard Shaw had in mind when he responded to evolution thus:

> … when its whole significance dawns on you, your heart sinks into a heap of sand within you. There is a hideous fatalism about it, a ghastly and damnable reduction of beauty and intelligence, of strength and purpose, of honour and aspiration.

Awkwardly, this is not a rational conclusion. Rather it is an emotional one. Being a good scientific model, evolution is purely predictive. It can tell us how physical relationships in the world are built, and from there can assist us in building up our expectations and our technology. What it can not do, however, is provide us with a compulsory mode of interpretation. For every person who sees evolution as highlighting the random, disinterested nature of the universe, there will be another who will prefer the words lucky and accommodating. For every person who says, 'See, God didn't make the world after all,' there will be another who responds, 'Ah, so that's how God did it.' And for every person who focuses on the reduction to matter and equations, there will be another drawn to the exquisite intricacy of the bigger picture. There is nothing strange or inconsistent in being both devoutly religious and a keen evolutionist. Evolution does not demand atheism, although it certainly makes it a more coherent stance for those who choose it. The big question of our existence, why there is something in the universe and not nothing, is still a long way from being solved. Many would claim we are still a long way from even framing the question which would make its solution possible. So to say evolution explains our existence is at best philosophically naïve, at worst wilfully misleading: a charge so often levelled at the religious side of this argument.

Evolution provides the most elegant description of part of the creation process, but that's all it does. The only metaphysical baggage it carries is that which we choose to load upon it. In this sense, it is not in principle any different from our theories on the birth of stars, the rotation of the planets or the pull of gravity. It's just another model of a mechanical relationship, to be incorporated into the meaningful picture we must all paint. It does not by its nature invite the sort of nihilism Shaw associates with it, in fact it does not invite any sort of attitude at all. The attitude is BYO.

There is a second class of error made by evolutionism's adherents which is worth considering. Just as evolution can not of itself provide us with our sense of meaning, nor can it in isolation lead to the building of our ethical models. This is the class of error Hume had in mind when he pointed out the move from 'is' to 'ought' is as common as it is foolhardy. Understanding the nature of the process that shaped our design and development will help define the boundaries of our existence; but with these fuzzy boundaries marked in, the drawing up of our aspirations is a matter for the story tellers. Again, not everybody gets this.

Consider just some of the 'should' statements that have been made in evolution's name. Some have claimed that evolution reminds us of our obligation to other animals. Because we were all shaped by the same process, they opine, because we too emerged from the slime, we should respect the earth's creatures as kindred miracles. Others however point out that evolution has shaped us to be competitive survivalists, out to extract resources and turn them into offspring. Through this contradictory reading, evolution not only sanctions, but also demands our exploitation of, the earth's life forms.

In the early days of the theory, Herbert Spencer evoked natural selection in support of his capitalistic theories, while over in the other trenches, the socialists argued evolution had produced our communal responses, and so the socialist system was the only one compatible with our scientifically-established nature. As already mentioned, believers and atheists alike have used evolution as the raw material for their metaphysical constructions, just as biologists today attack each other with vigour over the proper definition of race, claiming variously that their science shows race to be a biological or a social construction.

The existence of such opposing interpretations alerts us to the vital point that we are not in falsifiable territory. The human imagination is a wondrous thing, and has little difficulty twisting the theory to match the conclusions it finds most palatable. Finding the appropriate balance between the model and the story then requires a certain degree of honesty. We have nothing to fear from those who delight in spinning yarns to match their discoveries. We should be very afraid of those who deny this is what they are up to. This is hardly a new phenomenon. Indeed the abuse of evolutionary theory is as old as the theory itself. Enter into our story one Thomas Huxley.

Huxley was born into Victorian England, a culture quite different from our own. The Church of England was in the ascendancy; access to power, education and privilege were dependent upon one's class, and one's relationship to the church. A working welfare state was still a way off, and for those not lucky enough to be born into privilege, there was the constant threat of falling into the hole of poverty, never to escape. Church dogma was a powerful social force. Miracles occurred, prayers were answered; heaven, hell, angels, the works. The culture, with active support from the church, placed a premium upon tradition and propriety.

Yet things are rarely as simple as such generalisations assume. For on the other hand the great cities continued to grow, scientific discoveries accumulated, philosophers puzzled and pondered, class resentments fostered working class radicalism, and colonisation and commerce made globalisation a hot topic. Change was in the air.

Huxley himself was neither privileged nor dirt poor. His father was a school teacher, but the position was not lucrative enough to keep the family safe from deprivation (the more things change ...) Growing up, Huxley saw first-hand the dangers of the poverty trap. As a youth he moved to London, and got a job with a chemist, making deliveries through the city's slums. This was the London of Dickens: cess pits, child prostitutes, dead bodies left to rot in the streets. To a young man, the injustice and hypocrisy was too much to take. How could the church, with all its wealth and power, and its so-called commitment to good deeds, tolerate this state of affairs? One can imagine the adolescent rage that comes only from such a wellspring of self-righteousness.

Had Huxley been like most other adolescents, which is to say, had he in time tempered his rage, he would have been forgotten to history — but there was little chance of that. A talented academic, Huxley's way ahead was blocked by a lack of access to the resources of the privileged. He worked hard to win a scholarship, and became trained in medicine, getting his first taste of the world beneath the microscope. But having graduated and finding himself deeply in debt, he also found the establishment still blocked the way ahead. Thanks to his background, he would never have access to the lucrative professions of the upper classes, and instead would have to fight for the privilege to take on his first post: as a lowly paid ship's surgeon, the spectre of poverty still hanging over him.

On Huxley's first sea voyage, he spent his time aboard with a microscope, studying the details of jellyfish, determined to write the papers which would ensure his fame and status as a man of science. For, along with his rich vein of anger, Huxley also possessed vaulting ambition. The other legacy of his ocean voyages, beyond his rising scientific reputation and growing confidence, was the relationship that grew out of his meeting in Sydney with Nettie: his wife to be. That Huxley, who by his own account was fully smitten, would wait four long years before he was sure he was sufficiently financially stable to marry Nettie, speaks loudly of his great fear of poverty.

Huxley's interest in biological blueprints, and his great gift as a communicator, brought him into close contact with the most important biological theories and thinkers of the time. Once married, he and Nettie became close friends of the Darwins. It is perhaps not too cynical to suggest that in evolution, Huxley saw a weapon that could be brandished against his two greatest enemies: those members of the scientific establishment who stood between him and greatness, and the church. A lasting irony is that Huxley was himself a slow convert to evolution, and indeed throughout his life remained shy of the mechanism of natural selection. Huxley was perhaps less interested in the details of the theory, and more interested in the uses to which it could be put. For Huxley was a pugilist to the bone, a scrapper who could rarely resist the intellectual challenge of a duel.

Huxley has often been called Darwin's bulldog, the man who took the message out into the public, for Darwin was himself far too retiring to undertake such an aggressive campaign. I prefer the title of Darwin's spin doctor, for Huxley was well attuned to the needs and weaknesses of his audience. It is difficult, without recounting all the details, to give you a taste of Huxley's mania. He took control of journals and review sheets, lobbied his way onto influential boards, gave lectures up and down the country, published incessantly, reviewed with passion and vitriol, engaged in public debates and even set up a circuit of lectures for the working class, presenting evolution as an ideological weapon in their struggle against the ruling elite. He was a one-man lobby group, doing all he could to achieve his most cherished ends: furthering his personal status, establishing the professional credibility of scientific study, destroying the reputations of his greatest rivals, and bringing down the church.

It was in the last of these missions that Huxley was at his most unrestrained. His case against the church was a clever one.

Parts of the church, in ignorance of the workings of evolution, had aligned itself to a creationist story. Hardly the church's fault, in that prior to Darwin, all shared in this ignorance, and one could argue that even after Darwin, Huxley at times remained stubbornly blind. The true test of the church establishment was once again going to be its flexibility. How would it react to this new theory of evolution? It was, in essence, a Galileo moment, and again the church had plenty to lose. From the earliest records, human beings have used the evidence of intricate design in creation as an argument for the existence of a concerned, hands-on God, and as we have already seen, it is a compelling argument for the mystically inclined. So the church had good cause to promote the argument's power. If the world is the result of conscious design, and if the designer in question could be bothered to go to all that tremendous trouble, then it is only a small step to convincing people that this same creator loves them very much, and wants them to know him and follow his ways.

Once you've got that theology in place, then how better to maintain a certain level of power and privilege than by convincing the largely uneducated masses that you, the elite, have a mainline to God's thoughts? That's a particularly cynical rendering of the church's intentions at the time, I know, and rather overlooks the genuine devotion to good work within the institution. The point I am making is that no church in its right mind was going to give up the design argument without having a very good cause indeed. To admit that creating life as we know it may have been, well, no trouble at all, would be to embrace a massive reframing of theology. So while the church could have embraced evolution from the outset, it would have been surprising if it had done so. It was a high risk strategy for an established power, particularly if either the theory turned out to be wrong (and remember these were early, inconclusive days) or if the people were reluctant to embrace it (which still remains true today).

To the extent the church was slow to accommodate the scientific evidence (and this varied considerably throughout the church community, as you would expect) Huxley had a way in to the theological debate. The church stood for ignorance, Huxley suggested. By opposing knowledge, it was in denial of all that was good and true. And then Huxley took the familiar step. If the church could not be trusted to provide reliable stories, then nor could it be trusted to provide moral guidance. Indeed, Huxley managed to equate the two absolutely: only from a reliable model could a valuable story emerge. This may well be true, but as we have seen, it does not imply the next step of the argument, that from a reliable model a valuable story must emerge. This was the great flaw in Huxley's argument against the church and deserves closer examination.

Ironically enough, it was in Huxley's interest to ensure that the church took the reactionary rather than the progressive line. It would not serve him well for the church to see the error of its ways and embrace evolution. Huxley wanted the church to align itself with the weaker side of the argument, that he in turn may crush it. And if the church was sometimes too slow to oblige (for then as now it had its liberal intellectual elements), no matter. Huxley simply stepped into the breach, indulging in that time-honoured debating technique of setting up straw men that he might destroy them. While Huxley's debate with Bishop Wilberforce has now become the stuff of mythology, in fact it was only one such clash in a long series. I am not picking Huxley out just because his attitude was so extreme, but because he was at the very centre of the debate. Huxley did everything within his considerable power to make evolution the word on everybody's lips. And just look at what he was saying.

> [Science has] no intention of signing a treaty of peace with her old opponent, nor of being content with anything short of absolute victory and uncontrolled domination over the whole realm of the intellect.
>
> Blind faith is the unpardonable sin.
>
> Common principles of social action [are an] expression of natural laws.
>
> [...] the foundation of morality is to have done, once and for all, with lying; to give up pretending to believe that for which there is no evidence.

Huxley was setting up an opposition between science and religion, a battle from which he claimed only one winner could emerge. His key points, that faith is sinful, that social principles are extensions of natural laws, and that belief can rest only upon evidence, suggests that at this stage of his life at least he was confident that a rational, scientific stance could meet our moral needs.

It is not difficult to see how he might have been lured into such a position. Evolution was calling into question many of the details of the biblical text, just as the physical sciences had done in the preceding centuries. And the church, foolishly as it happens, was reluctant to back down on the points it was losing. Furthermore, it was not difficult to find church dictates which at first blush did not appear particularly moral, nor even consistent. Thus, concluded Huxley, the church has lost its moral authority. But now watch carefully the sleight of hand: because science has brought into question the church's moral authority, that hardly implies science must be the new authority on such matters. Yet it was implicit in much of Huxley's argument, and as he was someone who spent much of his career on the front foot, demolishing other peoples' cases, it wasn't an assumption he was often called upon to defend.

To be fair to Huxley, later in his career when his easier foes had been slaughtered, he did turn his considerable intellect to the harder problem. This was the 1880s, and the English economy was in a slump. In times of depression, it is not uncommon for people to look with hope towards prophets, and in this case both the promise of socialism and the magical claims of a laissez-faire economy were offered as ways out of the mire. As I've already mentioned, both sides were keen to enlist this new evolutionary theory as foot soldier in their war.

Huxley was drawn into this debate, and having himself more moderate political views, attempted to describe the relationship between the evolutionary 'is' and the social 'ought' in a paper entitled 'Evolution and Ethics'. And what did this old curmudgeon, scourge of the church, find himself concluding? That evolution could not in itself provide us with an ethical basis for our community. Indeed, he went so far as to suggest that humanity was called upon to oppose its own nature, in order to build the sort of society it yearned for. Where does such yearning come from? This was not made clear, and led some of his old enemies to conclude he had finally come over to the other side, accepting that beneath all this science there was a greater, spiritual compulsion guiding us. Of course, religious compulsion is not the only way out of this conundrum. It may well be that Huxley was staying truer to his sceptical roots, acknowledging the power of story to give shape to our deepest desires. But where Huxley in the end showed commendable intellectual restraint, others on both sides of the political debate continued to insist that moral lessons could be found hidden amongst science's box of tricks, adding an unfortunate veneer of scientific credibility to experiments both in eugenics and social engineering.

What Huxley's story shows most clearly is that from its very birth, evolutionary theory was presented as an opponent of religious thought, corralling people into an either/or dichotomy which continues to encourage the sort of simplistic thinking upon which all the fiercest arguments are built. And, in order to sustain the division, evolutionary supporters were forced to draw ethical conclusions from their new framework. While Huxley realised in time that such an approach was unsustainable, to a large extent the damage had already been done.

For scientific debates are often about anything but science. The truly tragic thing about the evolution sideshow is the vast amount of agreement it overlooks. For even amongst extreme Creationists, it is rare to find those who dispute the modern, predictive applications of evolutionary theory. Creationists do not argue in court that DNA evidence is not applicable to them, because they do not believe in these unique markers of accumulated mutation. They do not, as far as I know, argue that genetic engineering is impossible, on the grounds that nature is not constructed that way. Nor do they tell us we have nothing to fear from the bird flu, because everybody knows that viruses do not evolve.

On the important stuff, the battle has already been won. We agree on the basics of inheritance and the mechanisms of construction. In terms of prediction and application — that is, science — we all appear to be on the same side. In the end the debate is more about metaphysics, and that's not a debate science alone will ever solve. Sometimes you wonder if it wouldn't be better if we stopped sniping at one another and just got on with the exciting stuff. And in this, scientists, priding themselves as they do on their rationality, surely might be expected to stop rising to the bait and set the example.

Let me end this chapter with an analogy which I think will be helpful as we move onto my next argument. Consider an eager-to-please young father, who on the occasion of his son's fourth birthday, buys the lad a set of Lego building blocks. It is a fancy set indeed, with all the parts needed to make an elaborate medieval castle. Father's intention is clear. He and the son will spend some quality bonding time putting the castle together, with its turrets and drawbridges and little slots from which the defending archers (not supplied) can let loose their little plastic shafts of death. But son, being obstinately four, instead insists upon taking the various multi-coloured pieces out into the sandpit, where he sets about using them to make what in his eyes is a rather splendid spaceship. Poor dad is more than a little frustrated. No, he thinks, that's not what they're for. 'Can't you see? That bit's a drawbridge. Look, there's a little winch, to raise and lower it. It's just so, oh look, now what's he doing? He's chewing on the archer's head!' 'I'm making it into a space helmet,' the little boy innocently explains.

That pretty much is how I see science. It provides us with the pieces. And then we, the recipients of this fine gift, use our stories and our imaginations to make of those pieces what we will. Often the scientist has a fairly good idea how they would use the pieces. But like a good and loving dad, he or she needs to back off a little and allow us to find meaning and satisfaction where we will. The Lego pieces are real, solid things, and to argue they don't exist or aren't an important part of the final construction is batty. But equally batty is the insistence that they can only be used in the way their designer intended. 'If you're not making the picture on the box, you're not using them properly.' I think we're far too imaginative and wilfully contradictory to allow ourselves to be hemmed in by such anaemic thinking.

That's enough of the evolutionary debate: one that has spilled out of the scientific beaker and dampened the great unwashed. Now I want to explore one of the applications of evolutionary theory I hinted at earlier in this chapter, where the scientists are fighting amongst themselves. And I promise you they get rather vicious.

5 Gene, Genie

Many a dispute has been resolved by placing a wager and solemnly shaking hands. And many a dispute never will be. That's because there are two types of disputes. There are those which are based on a prediction, where both parties can agree upon the means of measurement and resolution, and those which are not. If I claim that gold is heavier than ice-cream, that there were only two series of the show *Carnivale* ever made (tragically) or that there are more white cars in the world than yellow ones, then anyone who disagreed could suggest we bet on it. We would enter into the challenge confident that such disputes can be resolved: by way of measurement, counting, or the discovery of counter examples.

Although it's a rough and ready sort of a definition, I like to think of facts this way: as things we are prepared to bet upon. This echoes Popper's observation that science deals in prediction, and hence the potential for falsification and progress. One interesting implication of this is that we know our 'facts' with varying degrees of certainty. For we ask ourselves not only, 'Would I bet on this?' but also 'How much?' I would perhaps be prepared to place my life on the fact that the sun won't rise in the west tomorrow, and if I owned a house I could be tempted to stake it against the proposition that the speed of light is constant (and there are physicists who'd bet against it). I'd bet my car on NZ's inflation being below ten per cent next year, but maybe only twenty dollars on the proposition that the same car won't break down some time in the next six months. I believe many things to be true in this way, while happily accepting some of them are more true than others. The shaky foundation of our facts need not concern us much. Predictive knowledge can still be wrestled with: resolved in the specific case and refined in the general one.

But what say I tell you that *Curb Your Enthusiasm* is the funniest television show ever made, or that sometimes it's okay to tell lies? We could argue about both these propositions long into the night, but we wouldn't take a bet on them. Instinctively, we understand that in matters of ethics and aesthetics, there is no simple way of adjudicating the dispute. Because no prediction is being made through these statements, no simple measurement will lead to both parties agreeing who ought to pay up.

This is not to say such issues should not be disputed and resolutions should not be sought; indeed, human society is hugely dependent upon nutting these things out. Which activities are to be encouraged, which made illegal? Which buildings are to be considered too monstrous to be built, which arts are to be deemed valuable enough to subsidise? What we must understand, however, is that the means of reaching agreement in these cases will be quite different from the blustering world of wagers. We have to find common ground first, and from there undertake a careful process of reasoned compromise, the end game being a position where at the very least all parties can agree to differ peacefully. The miraculous ability to resolve the immeasurable is one of humanity's finest achievements.

Because of this crucial difference between the way disputes are resolved in the world of science and the world of story, it is vital to know which realm we are operating in. Sometimes it's okay to say, 'Look, you're being an idiot, you're quite wrong. I'll show you.' (Stepping off buildings really does hurt.) Other times such intellectual colonialism betrays our own ignorance: the inability to see that we have moved away from testable science and on to personally constructed stories.

In the last chapter we saw how the evolution debate is inflated by such confusion. On the one hand are those who refuse to accept that their stories are directly contradicted by solid, testable propositions; on the other are those who infuse their models with meanings which the models themselves do not demand, and who refuse to accept this is what they are doing. But such disputes are not confined to tribal skirmishes between the scientific and non-scientific communities. Within science itself there are fantastically bloody and drawn out battles over the appropriate application of evolutionary theory. I want to examine the scientific credibility of evolutionary psychology, because it really is a compelling scrap. I'm also going to go out on a limb and pick a winner here. I think the evolutionary psychologists are backing a seriously shonky argument, which is a shame because it's built atop some stunning science.

Unless you've spent the last twenty years in splendid isolation, the rudiments of evolutionary psychology will be familiar to you. Indeed it's hard to open a newspaper without coming across another typically breathless report on some link between evolution and human behaviour. 'Scientists discover evolutionary link to obesity' (we need to move more and eat less). 'Biological foundation of taste in art discovered' (well, given that our eyes are biological …) 'Male reluctance to ask for directions linked to hunter gatherer past' (and we don't like asking for definitions either, so who cares what 'linked' means).

The public, and hence media, enthusiasm for this stuff is apparently boundless. We love being told why we are the way we are: it's one of our deepest fascinations. From astrology to Myers-Briggs, something deep within goes *aaahh* whenever we hear 'This is why you behave the way you do'. And of course the information being peddled is often profoundly comforting. 'Look, I'm sorry I'm crap at doing the dishes, but it's a proven scientific fact. Men don't notice dirt as much as women do.' Nothing goes down quite as smoothly and sweetly as the abdication of responsibility, which is why our bullshit detectors need to be twitching spastically at this point.

Whenever the message being pushed is the one people really, really want to hear, we need to subject the argument behind it to extra special scrutiny. So let's do exactly that. I want to explain a little of what I'll refer to here as the Evolutionary Hypothesis, the wonderful discoveries it's based upon, and to be fair, the areas where it does show some real promise. And then I want to show how, in the hands of the unwitting story teller, it can quickly become farcical. First a tiny bit of history.

The first seeds of evolutionary psychology were planted by Edward O. Wilson in the early 1970s, in his book *Sociobiology*. His suggestion that methods being used to deepen our understanding of animal behaviour could also be applied to understanding human behaviour, was at the time deeply provocative. So much so that people picketed his speaking engagements. That might seem a little extreme now, but at the time the social sciences paradigm tended to see human beings as anything but animals. Our behaviour, social scientists contended, was the product of our culture, and so only through studying this could we gain a proper understanding of how we came to be the way we are. The very idea that there might be such a thing as human nature, that we might arrive in the world with a set of genetically determined dispositions, was anathema to holders of the mainstream view. The fear was political as much as anything else. The social context was an era of protest movements: the struggle for racial equality, the battle to establish the tenets of feminism. There was an understandable concern that in the wrong hands this new 'science' would simply be used to dig deeper the trenches of our oldest prejudices.

There were some campus politics in play too, and much of the dispute did, and still does, have the whiff of turf war. Scientists were bullying their way into an area they didn't properly understand, claimed one side; social scientists weren't prepared to have their models held up to the rigour of scientific examination, claimed the other. So this was always going to be a fiery battle, the stakes raised by the underlying politics, the fury provided by the ever reliable tug of tribalism.

That this battle made its way out of the universities and became a staple of dinner party conversation owes a significant amount to another book from the 1970s: Richard Dawkins' *The Selfish Gene*. That sounds suspiciously like dustjacket puffery, I know, but the book has sold over a million copies and counting. That's the sort of achievement the rest of us can only grovel before, and is testimony both to the rare quality of the writing and the perspective-shifting message it carried. Dawkins took a hitherto academic subject and gave the rest of us a glimpse of its magic. In doing so he introduced us to a whole new way of thinking about animal behaviour, and how it might be put together. What follows won't make much sense without at least a rudimentary grasp of what's meant by genes, so let me give you a once-over-lightly treatment. Again, you should consult the bibliography if you want a more detailed, expert approach.

Contemplate for a moment the wonder of the cell, the building block of life as we know it. All the life forms you will be visually familiar with — grass, cows, human beings, worms — are made up of cells. The human brain alone has an estimated hundred billion cells, which I think you'll agree is plenty. Think of cells as tiny, specialised Lego blocks (when it comes to metaphors, I'm a committed recycler). These are put together to produce the miracle of a functioning life form. But, as the last chapter explained, there is an essential difference between an alligator made of cells, and another made of Lego blocks. When you see the Lego block alligator, no matter how convincing it may at first appear, you will be able to safely predict that somebody built it. It didn't just spontaneously shake together that way in the box. The wonder of life is that for the multicellular forms we see all about us, such self-construction does occur. Watermelon seeds grow into watermelon vines, which in time sprout watermelons with seeds in them, which in turn grow into watermelons. Which to me is sort of remarkable. How on earth does that happen?

At the simplest level, cells achieve the miracle by replicating, making copies of themselves. One becomes two, two become four, four become eight and so on. Only they are not, as development proceeds, exact copies. Your skin cells are different to your liver cells, or the cells in your brain. They specialise as they go. If they didn't, you and I would suffer from a distinct blobbishness of form.

To carry out this piece of replicative wizardry, a cell needs to be able to pull off four tricks. First, it needs to have within it the machinery for constructing a copy of itself. A cell must therefore contain, amongst other things, a little cell-making factory. Second, like any factory, it needs a system of energy and resource extraction. You don't get something for nothing, so it needs to be able to suck energy (sunlight for example) and chemical building blocks (nutrients) out of the environment. Third, it needs a recipe to follow. There has to be, within each cell, a store of information which dictates the way the factory goes about its task, so that the right sort of cells are made, and watermelon seeds don't grow into alligators (although that would be sort of cool). And fourth, there has to be a feedback system which allows for the fact that different sorts of cells need to be made during different parts of the production process, depending on which part of the life form is being worked on.

A replicating cell has all of this in place. As I say, here is not the place for the detail, but for the biological novice, the genome, the complete copy of the life form's DNA found in each and every cell, is the recipe part. It is a long string of molecules, which by dint of their shape, can be read off by other molecules within the cell (that is, these other molecules naturally form attachments with them). To carry on oversimplifying, DNA is made up of four 'letters' (distinct molecules), which can be arranged into three-letter 'words', which in turn can be arranged into 'sentences'. Each sentence, or gene, codes for a particular protein, or building block. Again the chemical properties of the molecules are exploited, with specific amino acids attaching to particular coded sequences, and strings of these amino acids in turn folding up into proteins, the building blocks from which the cell itself is constructed. Roughly.

As I've said, not every gene is at work in every cell. If they were, every cell would be the same. So a system exists to turn some genes on, while leaving others quiet. This works by way of a feedback mechanism, essentially communication from other cells and the outside world, to tell the factory where in the construction plan it is up to, or what particular environmental pressures it needs to be responding to. Again this is a chemical process, with some molecules able to bind onto stretches of the genome, stopping the stretches from being read. And as with words and sentences, it is not just the individual instructions which are important, but also the combination in which they are used. This combinatorial quality gives even a small number of genes almost unlimited scope for specifying different types of constructions.

So one way of thinking about your own development is tracing it all the way back to a little you-in-potential, a fertilised egg containing a mixture of genetic material from your mother and your father. Part of that information is used to guide the first cell replications, and from there a cascade of development is set in place, so dizzying in its complexity that it is capable of producing the miracle of you and I. The initial recipe then, your genome, plays a vital role in determining what sort of hair you have, how muscular your frame becomes, how your immune system functions, how susceptible to weight gain you are, and in some cases, even what you will die of (bad manners I know, to mention such things).

But that, as we all instinctively understand, is only part of the story. To take the example of muscularity, if you come from a particularly muscular family your chances of turning out that way are higher. But on top of that, how much you exercise, the sorts of sports you indulge in, the personal ambitions you develop and the range of drugs you are prepared to take, will all alter your physique significantly. Think again of a meal in a restaurant. The recipe followed by the chef, like the genes, has an important part to play in saying what the final meal will look like. But so too does the skill of the chef, the type of equipment she has available, the quality of the raw ingredients and the phone call that distracts her at the vital moment. So it is with cooking up a human being. You, the final product, are the result of a mixture of elements. That the particular package of genes you are born with is in some sense important is a point nobody disagrees on. Similarly, absolutely nobody is claiming those genes are the whole story. So far so good, then: nothing to argue with here.

The next step is to ask, where does the particular package of genes you are born with come from? This is a two part question, in that it asks both about the design specifications that go into making up the human being in general, and the particular take on human building that led to you in particular. The Evolutionary Hypothesis is interested in both these questions. Why are humans the way they are as a species, and what accounts for the intra-species variation? To answer these questions it is helpful to take Richard Dawkins' cue and ask, how is it that particular versions and combinations of genes persist through time? He urged us to think about this not in terms of the individual creature, but instead to look at the whole evolutionary process from the point of the view of an individual gene.

Let's return for a moment to our rat. Where does a rat get its genes from? Its parents. And what do the rat genes do? They provide the recipe for rat-like characteristics, both physical and behavioural. Now over rat history, a number of variants have been tried. We looked earlier at the idea of mutations, which are just alternative versions of genes, caused by copying 'errors'. Some may have caused longer legs, faster growing teeth, thicker hair, stunted whiskers or particularly twitchy reactions. The question the selfish gene perspective asks is: why have the genes currently washing about in the rat population succeeded, where others have failed to stay in the mix? Why do rats, in other words, look and behave as modern rats do, rather than any of the other possibilities that might have emerged?

Part of the answer, as we have already seen, is blind luck. Who knows what nascent rat innovations were lost in fires, drowned in floods or buried beneath landslides? But the other part of the explanation can be found in the inexorable logic of natural selection. Genes that survive in the population, Dawkins reminds us, are those that behave most selfishly. By this, he doesn't mean genes are independently motivated players, but only that they behave as if they were, in the same way that water molecules in a river behave as if they are dead set on heading out to the ocean.

The gene's apparent motivation is to promote its own reproductive success. Many have criticised the use of the word selfish in the phrase, given that consistency, rather than selfishness, is the salient point, but we'll leave the semantics to one side and go for the gist of the thing. Imagine two mutations which each produce behavioural variations in rats. One makes a rat more inclined to work co-operatively with other rats when scavenging for food. Another makes a rat more inclined to shun the company of other rats, scavenge alone, and when it encounters a competitor, fight to the death to defend its food supply. The pertinent question becomes which of the two mutations will spread through the population? Which, over time, will become the standard rat configuration, and hence determine typical rat behaviour?

The answer is brutally simple, and simply brutal. Whichever mutations lead to a rat leaving more offspring on average over the course of its life are the mutations that will spread. This of course will depend on factors like resource availability, the rat's physical profile, the behaviour of other rats, and the like. Two things it will not depend upon, despite their intuitive appeal, are what is good for the rat population as a whole, and what seems cute and cuddly to Walt Disney storyliners.

The criteria of the Natural Selection Design Unit is simple-minded and consistent. Will this innovation lead to higher reproductive rates for those individuals who possess it? Because reproduction is the means by which new genetic profiles spread through populations, this is the only 'rule' available. So, if you look at the genetic profile of any creature, it is fair to assume, blind chance aside, that this profile is the result of its genes having outcompeted other genetic versions in the past. And yes, in a sense this is blindingly obvious and uninteresting, but what we're working towards here is a perspective shift. I am still struck by the power of the oft repeated observation that every one of us is the last in a long line of consistently successful reproducers. Not a single one of our ancestors was too shy, weak, slow-witted, ugly, stupid or unlucky to find a mate and oversee the successful arrival into the world of the next generation. Not a single one. Success really does breed success. And so it is the genes which aided that success, that acted with apparent reproductive selfishness, which made it into the contemporary mix.

It is easy for us to link genes to physical appearance, because it is by this means that we naturally categorise the living world. A cat in this respect is like a good piece of art: we know it when we see it. But in the animal kingdom it isn't just the physical profile which defines the creature. It is easy to overlook the fact that each animal also has a typical set of behaviours, and 'selfish gene thinking' helps explain why. A young bird might have a go, with encouragement from mum, at stepping out off a cliff top, but a young donkey is less likely to. Cows don't throw themselves into fast-flowing rivers when they're feeling hot, and cats don't leap into fires because they find the flames bewitching. Maybe some mutants in the past tried such behaviours, but if they did those mutations were subsequently lost to evolutionary history, for obvious reasons. They were not successful reproducers. So selection pressures herd animal behaviour down those paths which will lead to ongoing reproductive success.

And again, there isn't anything too controversial here. Although debate rages over terminology, and there's been an ongoing war over the appropriateness of the 'gene' as the unit of selection, when the chalk has settled on the academic head-butting, the disagreement does not, within this context at least, appear substantive. All agree that selection pressure has led to much of what we think of as animal behaviour, even if the pathways may have been far messier and more indirect than simple examples suggest. It's just this sort of messiness that is driving the next wave of research, which is exactly as it should be.

What's more, the sort of testability demanded by Popper can be met at this point. For the theory predicts both that specific genes will lead to particular physical developments in the brain, and that these developments will in turn impact upon the animal's behaviour. Thanks to our ability to modify genes, rigorous experiments can be carried out in this field, and that's exactly what's happening. So if we're looking to derail the hypothesis, here's not the place to launch an attack.

The next step in the argument is the interesting one, and is essentially the leap Wilson encouraged us to make, and many others still beg us to resist. What if the human brain, an exquisite example of a carefully honed biological device, has been shaped by the same selective forces that taught dogs to growl and cats to plead? What if the behaviours we indulge in are less a matter of choice or social conditioning, and more a matter of the dictates of our evolutionary past?

There are two characteristics of the human brain which make this a very interesting thought. The first is that human brains have a tendency to reshape their environments. Think motorways, microwaves or nightclubs. Brain evolution happens slowly, it takes many generations for a new mutation to spread through a population and it's reasonable to expect that the human brain today is not substantially different in its design than it was say twenty thousand years ago. It is far less reasonable of course to assume the same about the environment. Take the lifestyle of a citizen of a major world city like Tokyo and consider how the survival pressures differ from that of our nomadic, hunter gatherer forebears. But if the modern brain was slowly shaped by selective pressures in this far-off environment, it may well be that many of the apparently irrational behaviours we struggle to shake today owe something to this mismatch. I for instance have a significant part of my brain dedicated to a fear of snakes. The bloody things terrify me, which, given I've grown up in a country without them, does seem a little odd. Meanwhile potential killers that should worry me much more, fast food for instance, don't frighten me at all. (In fact I find it distinctly pleasant.) Perhaps then the story of the modern human is that of a stone age brain trying desperately to cope with a space age environment. It's certainly an alluring thought, and sits at the heart of the Evolutionary Hypothesis.

But there is another possibility. What if the human brain design selected by evolution put a premium not on stock behaviours, but on plasticity? Perhaps, in general, the brains that were best able to respond to change by inventing new behavioural patterns for themselves did better than those which relied too heavily upon narrow ranges of thinking. Perhaps the very thing that sets us apart from the other animals is our ability to constantly reinvent ourselves in the face of new challenges. Perhaps that's the secret power the arrival of language-based culture unlocked.

Of course both these things can be true at once. Our culture may give us our flexibility while our biology sets the parameters. Indeed, put like that, it sounds perilously close to a truism. The argument of course is about the detail. Which of our behaviours are easiest to change, and what are the best methods of changing them? Are women in general less interested in sport because our culture tells them to be, or is there something about the female brain that limits the interest? And why am I so useless at keeping in touch with people and remembering birthdays? Can evolutionary thinking help with this sort of question? As I've already hinted, I think it's at this critical step that the whole hypothesis starts to wobble, but let's give it a decent chance to clamber up onto its pedestal before we start heaving things at it.

I'll stick with the reliably explosive issue of gender difference. Here's an area where everybody, it seems, has an opinion, a fact well-exploited by publishers and talkshow hosts. It's an area where the practical implications of the debate are plain to see. How do we create a society where women are safe from sexual assault? How do we give both boys and girls a chance to engage fully with education? How do we make sure that both sexes are heard when it comes to decision-making? The extent to which culture gives us the flexibility to modify definitions of gender is important not just as a prompt for dinner party conversation, but also as an influence on key policy decisions.

Luckily enough, it's also an area where evolutionists feel they have plenty to say. So how do they do it? One set of clues comes to us from studying other members of the animal kingdom. By carefully watching birds, mice, chimpanzees, etc., naturalists have noted all sorts of differences in gender behaviour. Why do birds, which are primarily long-term pairers, cheat on one another? (Notice how quickly we lapse into outrageously human interpretations here.) Why do certain male creatures insist upon banging heads together in an attempt to win over a mate? What is the peacock doing with such a ridiculous tail, and why will a bee sacrifice its own life in order to defend the hive? The theory, teased out by biologists in the latter part of the last century, relates it all back to the animals doing whatever is best for the survival chances of the genes they carry. Those who didn't follow such strategies in the past didn't pass their genes on. There is a compelling sort of a logic to this. (And how does a bee killing itself promote its own genes? Ah, glad you asked. Male bees have only one copy of their chromosomes, females two, and due to this quirk, a worker bee is more closely genetically related to its sisters than it would be to its own offspring, and this has interesting mathematical implications for the way selection does its work!)

Given that animals behave in whatever way will maximise their chances of passing on their genes, we might expect males and females to adopt different behaviour strategies. For, at least in the animal kingdom, the reproductive potential of the two genders is vastly different. A typical male can leave many, many offspring, just so long as he can find enough willing mates, and fight off competitors in the process. A female, by contrast, is constrained by the fact that she has to invest time and nutritional resources into the offspring. A female bird can only hatch so many eggs over its lifetime, far fewer than a male bird can fertilise. (Quaint isn't it, all this talk of birds and bees?) The argument says female animals will have developed strategies to maximise the survival chances of its offspring, perhaps by choosing a genetically well-endowed mate, or one that will protect and provide. It's interesting to note here that mothers have sons as well as daughters. So, if Mrs Bird manages to score a particularly unfaithful, competitive, attractive and sneaky Mr Bird, then her sons are more likely to inherit some of these characteristics, which will increase her genes' chances of spreading! Here is a potential reason why the good-looking male (he of the impressive tail perhaps) gets all the birds. Maybe the hens are hoping for good-looking children, who will repeat the trick, all in the name of their genes.

'But hang on a moment,' you complain. 'Birds don't "want" their genes to succeed. You can't talk of birds "hoping" for things. That's just humans projecting their storytelling needs isn't it? And even if it wasn't, you can't possibly apply it to humans, because we certainly do not go through life rationally calculating the impact our behaviour will have on our genetic legacy.'

Both these objections are reasonable, and yet oddly both sides in this debate agree they are irrelevant. This comes down to the difference between proximate and ultimate cause, and is worth dwelling on, to avoid later confusion.

A bird does not pursue gene maximising strategies in order to leave more genes in the future. Rather, it pursues these strategies because it has been programmed to do so: programmed by the genes of those who followed these strategies in the past, and who in doing so came to dominate the gene pool.

Think of it in human terms. Imagine that in the past a certain type of man, let's say a particularly violent, jealous, fiery type, scared off many of his male competitors. He may not have done this in order to increase his reproductive chances. He might just have got a kick out of being a prick. Nevertheless, so long as improved reproductive outcomes were a side effect of such behaviour, the next generation contained a higher proportion of violent, aggressive males. And so, over time, new behaviours spread, and we end up behaving as if we were motivated by genetic inheritance, even though such a thought may never pass through our minds. Such is the theory.

Now it takes another interesting turn. Some time in the past our human brains grew, and perhaps then, perhaps later, began to acquire the thinking characteristics of the modern Homo sapiens. There is a lot hidden in a word like grow, so we need to be aware of just how intricate an organ the brain is. We're not just talking about a bigger brain, but also about the sort of detailed design specifications which for now still elude our deeper understanding. This average brain growth was the result of changes in the genetic profile spreading through the population. How did they spread? Presumably on the back of differentiated reproductive rates: our old friend natural selection. Brain designs which gave relative survival and reproductive advantage are those which lasted.

What can this tell us about gender roles? Well, one thing we know about the human being is that it requires a huge amount of resources to bring one up. Watch a newly born lamb. In no time it is up on its feet, following Mum around and bleating for a feed. By comparison the newly born human is relatively helpless. There is a good reason for this: it goes right back to the big brain thing. Unlike the brains of other mammals, ours do a lot of their growing post-birth. It's the price you pay for being a fat head; if it did any more growing in advance of its big debut, birth would become physically impossible. So, in order to bring up a little human, someone has to invest a lot of time and energy into looking after the newborn. This, along with the already significant demands of a long pregnancy, means that back in the resource-poor developmental past, Mum was expending a lot of her precious energy on bringing up the next generation.

Put our situation like that, and big brains start to sound like a distinct disadvantage. Not only do they force us to be born helpless, and force adults to devote massive amounts of their own time protecting and nurturing us, but they also leave us with a big weight hovering over a thin neck, making us ridiculously vulnerable to injury throughout our lives. Yet brains did indeed grow bigger, and the series of mutations supporting this new innovation spread. There must have been some huge compensating advantage. Big brains must have made us better at passing on our genes.

This sort of evolutionary reasoning leads us to speculate about the strategies that might have made males and females better at winning the reproductive war. Might these behaviours have been programmed into our expanding brains? It also allows us to wonder if male and female human brains might have developed a different bag of tricks to meet their separate pressures, as has been the case with many other animals.

Those women who were successful at bringing up children (those whose genes we ended up with) must have developed certain strategies: strategies which, if genetically conveyed, may underpin modern female behaviour. These were not necessarily conscious strategies: it's just that women who naturally followed them left more genes behind. Suggestions from modern biologists of what these strategies were include the ability to discriminate between healthy and unhealthy mates, and a tendency to choose powerful mates, who have access to vital resources. Choosing sneaky mates, who will leave sneaky sons to continue the cuckolding tradition is also a possibility. So is the development of strong social networks, in order to share the child-rearing task. (Awfully, it is suggested a strategy employed by some female birds is to find the most reliable, dependable male, and then whip off for a secret tryst with the highest status male available. The best genes nurtured by the most reliable dolt. Oh dear.)

If the key female trick is keeping the little ones alive (those who were bad at this, remember, will have been filtered out of the population long ago) the male trick is to father as many little ones as possible. The assumption here is that for the male there is a trade-off between devoting resources to ensuring a particular child survives, and spending time and energy on ensuring many such little children are born in the first place. With females, the trade-off is weighted in favour of nurturing because of the upper limit on the total number of children they can have. So selection may well have favoured men who were good at getting hold of those things which females valued, like resources. Just as it may have favoured males who fight off potential competitors. Cue ambitious, materialistic, competitive, aggressive behaviour, comparatively speaking.

Again the theorists look to the animal world for back-up. If there is any merit to these sorts of suggestions, then differences in gender behaviour should link nicely with differences in the resources needed to bring up a child. So birds, for example, which must devote a lot of time to looking after an egg before it hatches, face a situation which is broadly analogous with the human problem. Interestingly, birds, unlike most mammals, tend towards monogamous relationships.

Selection has favoured the fellow who sticks around. Hence the ability of the film industry to interpret penguin relationships in such outrageously human terms.

By taking an evolutionary perspective we are led to expect behavioural differences between the genders: differences which are in some sense hard-wired genetic impulses leading to different physical structures within the brains, and hence different ways of thinking and acting. This suspicion, based as it is upon an observation of the way selection pressures have played out in other species, and upon some deductive reasoning about the particular constraints faced by the nascent human race, becomes a sort of hypothesis machine through which evolutionary psychology defines a research programme.

In relation to gender difference, the research programme does look at least feasible. Modern imaging techniques allow researchers to examine which areas of the brain are doing most work when particular tasks are required of them, and it does seem to be true that males and females are typically using different parts of their brains to complete certain tasks. Furthermore closer examinations of brains show there are average structural differences between male and female brains. Links between brain development and other gender specific development paths, e.g. testosterone production, can then be looked at, and from here there is a chance of getting a firmer idea of the way nature and nurture work together to achieve the gendering of the brain. And the better transmission mechanisms are understood, the better we are able to see the way certain behaviours and reactions can be both triggered and moderated. In terms of education for example, an understanding of gender differences in the development of language capacities may lead to important discoveries regarding the way the literacy gap might be closed.

I appear to be contradicting myself. For I said at the outset that I didn't much buy the evolutionary psychology argument, yet here in outline is a programme that does appear classically scientific. Existing theories regarding natural selection are used to generate new hypotheses which in turn can lead to the predictive work required of any genuine research programme. So where's the problem?

Well, I've done what many commentators in this area do: skimmed over the dodgy bits. I did this to get the basics of the theory down. Now you have an idea of what's going on, let's go back and expose some of the storytelling.

The first, and perhaps least damaging appearance of the storytelling tendency, occurs in the many field studies where animal behaviour is observed and recorded. It is very easy, when skipping over a sentence such as — 'The female birds are shown to be more likely to have affairs with the higher status males in the colony' — to imagine that the method used for reaching this conclusion was all very scientific and clean of subjectivity. But let's just think about it a little. Someone, somewhere, has come up with a definition of highest status male here, a loaded term if ever there was one. How was this measured? Are we talking about the birds that are most aggressive towards other birds? Win the most fights? Eat the most? Or just get the most girls? (Here the reasoning becomes dangerously circular, for we would be noting that the females are more likely to mate with the males who mate with the most females … well, yes.) Whatever the methodology, it's worth remembering that what we have here is some poor shmuck with binoculars and an anorak perving at birds and scribbling in a notebook. We humans are notoriously good at seeing what we want to see, and attributing human motivations to very non-human scenarios. Whatever it was that motivated the particular researcher into this quaint area of study, it will have something to do with his or her view of the world. And that world view can very easily leak back into the observations. So while it would be foolish to suggest that all field studies of animals are unreliable, it would be equally foolish not to expose some of the more spectacular claims to fairly serious scrutiny.

More important is the easy assumption that what is true of animals is also true of humans: what I think of as the Disney approach to animal study. As an example, Frans de Waal is an expert in primate behaviour and has written extensively and with much authority on the subject. In a recent fascinating essay entitled 'How Animals Do

Business' he described experiments which sought to test the way chimpanzees paid one another for participating in co-operative tasks, and placed value upon their rewards. For example, chimps were taught to exchange pebbles they had been given for food, namely slices of cucumber. Chimps apparently had no trouble with this concept of exchange. Later, though, a new reward was introduced. Some chimps were offered not cucumber for their pebbles, but rather grapes, a treat far more prized by the primates. And here the response became interesting. The grape receivers were chuffed of course; but the others, seeing that some of the group were given grapes while they got nothing but cucumber, were reportedly miffed, throwing tantrums and suddenly no longer being satisfied with their cucumber payment.

That's a cool little experiment, and it does tell us something sort of cute about chimpanzees. They seem to value the relativity of rewards, much as we do. No problem with that. But the next step is, I think, a rather remarkable one. For the article suggests that economists need to take this evidence into account when building their models of human behaviour. The implication is that because relativity measures are a standard part of chimp response, and because we share a great deal of chimp heritage, we should expect relativity to be a core part of the natural human response.

The two errors of reasoning here are plain to see. First, why on earth study a chimp to learn about human behaviour when there are perfectly good human subjects to observe? Yes, we may share this base characteristic with chimps, but then again we may not. Our brains and our genetic heritage are clearly similar and clearly different. The only way to see if this behaviour is also present in humans is to test humans. The chimp result adds no credence to the conclusion, one way or the other. We already know humans are jealous, competitive creatures: we have thousands of years of story attesting to this fact, and in fact it is exactly this knowledge which makes the chimp result so arresting. Look, we think, they're just like us. Studying animals doesn't tell us anything new about human behaviour. It just tells us new things about the animals, of course.

The evolutionary enthusiast may argue: ah yes, but because it is present in both chimps and humans, this tells us it is a fundamental part of our genetic heritage. Of course, by the same reasoning, so are our legs, and again I don't need to know that chimps have legs in order to be certain that they are a key part of the human design.

For human beings quite clearly are not exactly like other animals. Amongst other differences, when we head out in the field to conduct our studies, we do not find other species sitting there with notebooks, collecting their own data. So, while animal studies might give us a few clues regarding how evolutionary pressures assert themselves in general, what we actually need is an idea of how they all worked in the specific case of humans. We need to get to the heart of what defines the human creature — things like complex language, abstract thought, and sophisticated cultural transmission mechanisms — and work out what the selection pressures were like during their evolution.

That would be fine, if we knew anything much about any of these things, but we don't. The truth is we don't have a whole heap more to go on than a few fragments of bone, some cave paintings and the non-perishable remnants of the odd ancient burial site. This is not to belittle archaeology: what we have found is quite incredible, and some of the piecing together of the clues has been heroic. But the absolutely key stuff — who went where when, in what numbers, under what climatic conditions, with what brains, working in what types of communities, developing what languages, for what particular purposes, etc. etc. — is still largely unknown. When were the first words uttered? What were they used for? When did we first become aware of our own mortality? When did the modern sense of self emerge? These are important questions which simply must be answered first. It's not credible to use our evolutionary past as the basis for a psychological hypothesis if we don't know exactly what that past was like.

Which leads to a most interesting point. For evolutionary psychologists don't use this past as the basis at all. Rather they pull a great big cheat on us, and hope nobody will notice. Let's rewind and consider an example I dropped in earlier. Selective pressures encouraged males to adopt behaviours leading to the liberal spreading about of their genes, I speculated; while females specialised in ensuring the survival of their offspring. And from here, as is the standard practice, I painted a picture of successful women being those who were particularly good at forming networks, attracting high status males, judging the health of potential fathers, and so forth. The male by contrast was the aggressive competitor, subject to jealous rages, ambitious to a fault, seeking status and domination of resources.

The trouble with this picture is that it's nothing more than a story, an entirely arbitrary imagining of our evolutionary past. To make the point, ask yourself what's wrong with the following re-interpretation. Because women needed to ensure the survival of their offspring, they needed access to the resources of a providing male. So competition for the best providers became fierce amongst the women and selection favoured violent, aggressive, and fearfully jealous, ambitious women. Meanwhile, as our brains developed to cope with complex strategising, men soon realised that the best way to out-compete potential cuckolders was not to indulge in the traditional mammalian head butting, but instead to lay on the charm. For while Conan was out front smiting all-comers with the jaw bone of an ass, Romeo was out back getting intimate with the womenfolk. So as complex culture evolved, selective pressures led to gentle, smart and sneaky males: leaving the angry oafs mateless.

Two entirely plausible and directly contradictory hypotheses can emerge from the Evolutionary Hypothesis. No problem, you say. We can test them. We can turn to studies of contemporary humans, define a characteristic like aggression (tremendously hard to do, by the way) and then measure it in males and females. This will in turn lend greater support to one of the two evolutionary stories. And of course that's exactly how researchers proceed.

But the problem will by now be screamingly obvious. The whole thing has become entirely post hoc, the method Popper cautioned against. We are not using evolutionary theory to shed light on modern psychology at all. Rather we are using modern psychology to shed light on evolutionary conditions. Shoot first, paint the target later. And there's absolutely nothing scientific about that. The evolutionary part of evolutionary psychology becomes redundant and we are left with the crashingly unoriginal observation that the way to understand human behaviour is to study human beings, which social scientists already knew well before the biologists arrived armed with their studies of the peacock courting ritual.

What we have to be very careful about then is the modern fashion for tacking a hastily concocted evolutionary explanation to the back end of a piece of research in order to give it greater credibility. This is particularly problematic if the research was prompted by an existing prejudice, as is often the case in the area of gender. So next time you read that scientists have come up with a new and wonderful evolutionary explanation for differences in female and male attitudes towards sex, remind yourself that the quality of the evolutionary story is not the criteria by which the research should be judged. The story is a post hoc add-on designed to make the research more sexy: a sort of scientific peacock's tail, to stick with a theme. Whether indeed there are gender differences in this field can only be resolved by observing people in the here and now. Ignore the 'long grass in the savannah made copulating couples vulnerable to surprise attack, and led to a male preference for being on top, to facilitate an easy escape' part of the argument, and judge instead the quality of the modern survey being used to support the theory. (As an aside, do be suspicious of those surveys which show that on average, men have more sexual partners than women. It's mathematically problematic, you see.)

We shouldn't dismiss the Evolutionary Hypothesis altogether, but we should require it to reach the same standards we ask of other endeavours. When biologists do indeed have such a good grasp of the evolutionary conditions under which modern humanity emerged that they are able to generate information-rich hypotheses which make unique, surprising and testable predictions about modern human behaviour, there will be such a thing as evolutionary psychology.

Remember, nobody expected mass to bend light before Einstein, and nobody expected prisms to be able to make white light of rainbows before Newton; but evolutionary psychology yields no predictions which can't be found in Shakespeare. The difference between prediction and story is, as ever, vital.

It's tempting to leave the whole nature/nurture debate behind with a final dismissive flourish, but that would overlook another fascinating aspect of this debate. For irrespective of the narrative abuses that plague this area, the evolutionary perspective does introduce one vital element into the psychology debate: the link between the genes we carry and the way our brains work. No matter what precise forces in the past led us to having the brains we have (and in falsifiable terms, this appears for now unknowable) it is still interesting to wonder to what extent these brains are flexible, and to what extent they are dependent upon instinctive patterns of thought. Just because a lot of the evolutionary stories about gender are silly, doesn't mean there are no evolved differences between male and female brains. If these differences are sufficiently robust that culture will struggle to change them (will dressing your little boy in pink make him gentle, or just piss him off?) we ought to know about it.

As I've already said, one of the startling characteristics of the human mind is its plasticity. You might have a tendency to get angry quickly, take unnecessary risks or disagree with everything you hear on principle, but you also have a capacity to moderate all of these things. Sometimes just by pulling yourself up and saying, 'No, don't be a dick,' and other times by undergoing extensive periods of therapy. Either way our biology is hardly our destiny (and neither side says it is). The information we are after (for instance, what would be really, really useful to a school teacher grappling with the topical issue of how to get boys motivated in class without compromising any of the hard-won gains made by girls over the last thirty years) is which instincts are easy to change, and which are hard? Let's say we know we want boys to be co-operative, gentle, sensitive and eager to learn: that this is a story our culture has agreed upon. Is there anything within the evolutionary hypothesis that can provide us with a little practical advice on how to achieve this, or on whether the aim is realistic?

There are two ways of going here. One of them is, I think, misguided, and the other potentially interesting. I'll start with the misguided one. Go back to our chef: this time she is undergoing a training programme. The tutors will no doubt highlight a number of important things. How to choose the best ingredients for the dish, how to store it, handle it, prepare it, combine it, cook it, etc. They will talk about timing, about knife selection, about temperatures and about presentation. If you were to ask the tutors where this knowledge came from, they would say it was through traditional knowledge, which was augmented by experience and experimentation at each stage of the communication chain. Wisdom by way of an elaborate process of trial and error.

At no stage, I am prepared to bet, are the budding chefs going to be encouraged to take all the elements of food preparation and put them into two arbitrary camps — let's say ingredients versus process — and then indulge in long and pointless arguments about which is more important. Nor is any fool going to undertake a scientific study to show what percentage of final meal quality is attributable to any given element. No one is going to argue there's no point learning how to cook, because poor ingredients can't be made tasty, and nor are they going to suggest that we don't bother discriminating between good and poor ingredients, for the well-trained chef can do miracles. Chefs, as a general rule, are just not that stupid. If asked, they might explain that such an approach is ridiculous, for food preparation is too complex, too subtle, too instinctive, to lend itself to that type of reductionism. So how about raising a child?

At least one branch of study in this field has tried just such a quantification trick in the field of human development. Researchers, determined to show once and for all the impact of our genes on our upbringing, undertook the now famous twin studies. The thinking behind the twin study is initially seductive. Take two identical twins, who share all their DNA, and compare their development paths to non-identical twins, who despite also being raised together, share on average only half their genes. Find measurable things, like IQ, and see whether over a large study the identical twin scores are more convergent than those of non-identical twins. This presumably would be evidence of a genetic effect. The same sort of study can be modified in a number of ways. Compare identical twins raised apart to those raised together, for instance. Do the same for non-identical twins, by way of a control. Large scale adoption programmes, well documented in some countries, have allowed this sort of work to be done.

What have these studies found? A number of points are made, some purely anecdotal, others more carefully quantified. On the anecdotal front, much is often made of two identical twins raised apart, who had never met, yet when collected separately from the airport both remarked that the car's wheel bearings needed attending to. Spooky huh? Well, not to a sceptic. Dodgy bearings make a distinctive noise. Lots of men invited to sit in the back seat, directly above such a problem, might comment.

On the quantified front though, something more interesting happens. We are told things like the variance in identical twin IQ is lower than that for non-identical twins, and that this effect appears to be largely independent of whether the twins are raised separately or together. From this the numbers are crunched. We are told that IQ is fifty per cent hereditary for instance, and perhaps only ten or fifteen per cent due to immediate environment. Similar results come through for things like personality traits. Yes, I know. How do you measure that? That's another argument altogether and I'm going to ignore it for now, because even if it turns out the measures are reliable, there's a deeper mathematical problem to consider.

Initially the twin study results do appear a little stunning. 'So what, upbringing doesn't affect IQ? But what about all the baroque music CDs I played my baby in the womb? Darn, I was ripped off!' Well yes, in all likelihood you were, but most of us work that out without the studies. As with any research, it's always worth calming down a bit and peering behind the numbers. In this case there's an awfully big gap between what the numbers tell us, and how the commentators sometimes use them. I blame maths teachers. We're just not getting through to you guys.

One thing the studies can reasonably claim to do is show that genes do have an influence over things like personality and intellectual development. We'd be fairly shocked if they didn't. In fact it would pretty much blow the entire genetic development hypothesis out of the water and biology would be in panic mode. Anybody who resolutely believed that inheritance had no impact on how we turned out, that we are all born blank slates, would be shocked by twin studies. The verdict is in, folks: genes and environment work together on this one. But back on planet sanity, it's going to take something more to impress us. That something is provided in the form of summary statistics: attempts to measure the relative importance of the environmental and genetic inputs. Sadly though, that bit is less compelling. This is where the chefs start laughing at us.

Think about it this way. Take sets of identical twins raised apart, and examine the impact the environmental difference has by comparing them with sets of identical twins raised together. Okay, so far so good. But wait a minute, how will those environments differ, apart from not having identical twins in them? If one twin is lavished with encouragement and attention, and the other is locked in a dark room and never talked to, then I am prepared to bet that the environmental effect will swamp any genetic influence. Similarly, raise them in environments which have similar routines, levels of support, etc., and the genetic signal will dominate, simply because the environmental signal has been muted. It's like arguing whether driver skill or car quality is more important in determining the outcome of a car race. It rather depends upon the cars, and the drivers. Put very similarly skilled drivers in identical cars, and skill (and luck) will determine the outcome. Put them in wildly different cars, and the driver influence becomes negligible.

Those reporting on twin studies usually point out their results only refer to the variance of a characteristic found within a given population, but they often say this as if it is a minor point. And then they undo the good work by drawing the oddest conclusions about the relative importance of genes and environment. It's not my intention here to name names. Some do this analysis well, but most are slipshod and overexcited. Check out some of the literature for yourself and try to spot the difference.

To reinforce this, and also to show a way in which such studies might be informative, consider a hypothetical result that showed the difference between identical twins' grades was not much affected by whether they were taught in the same school or in different schools. It would be foolhardy to then claim that genes play an important role in determining IQ, whereas schools do not. It just may be that most of the schools covered by this survey were doing the job pretty well and hence the choice of school made no difference. That's a world away from saying schools make no difference, or from concluding that genes are more important than schooling in determining educational outcomes. (That is again a meaningless statement. Send your child to a school where the students are not allowed to speak to one another and have no access to books, and schooling makes all the difference.)

But, if we let our eyes slide off the silly nature/nurture debate for a moment, the study can show up interesting things. If all the schools are doing a roughly similar job, despite the fact that some have uniforms and some do not, some allow eating in class and some do not, some are single sex and some are not, this gives us really interesting information about the sorts of things which may or may not tend to knock students off their developmental paths. It tells us something about robustness. In an age where there is a lucrative industry pandering to the paranoid parent market (sign language to enhance brain development, finger exercises for a more co-ordinated baby, leek soup for stoicism ...) it is good to be reminded that children seem to be able to cope with a wide range of developmental environments. Indeed, this may be one of the more important contributions the Evolutionary Hypothesis has made to the human nature debate.

But the next step, to argue that genes are important to developmental outcomes, and environment is not, is a step too far. We're measuring something discrete — genetic code — against something nebulous — environment — and we should therefore be very suspicious of any answer that comes out as a number.

There is an alternative to twin studies though, when it comes to examining the link between genes and behaviour, and that's the longitudinal study. The most famous of these has now been going for over thirty years. An entire cohort of babies born in a Dunedin hospital has been followed up with extensive testing and interviewing over their lifetimes. Once the technology became available, research into DNA sequences became possible. Such an approach makes it feasible to define subgroups: for instance those who got in trouble with the law, and then to search for specific genetic correlations. Maybe, for example, a particular genetic mutation might associate with a high tendency to indulge in risk-taking behaviour. Where a particular aspect of brain function is sufficiently well understood, and some genetic links to this function are already known, the search can be narrowed down to such a degree that tentative conclusions are possible.

What has this sort of research shown? There's evidence that specific genomic variations can increase the likelihood of particular behavioural outcomes (delinquency for example) given the right environmental triggers. So, a person with genetic variant A, may be no more likely than B to get in trouble in one environment, but may be twice as likely if they are both exposed to an alternative lifestyle. The same sort of thing has already been established elsewhere with physical health effects. A person may have a gene making them particularly susceptible to lung cancer if they smoke, which apparently has no impact on their health at all if they are a non-smoker. One suggestion coming out of the Dunedin study is that a certain type of youngster (potentially genetically identifiable, which raises all sorts of interesting ethical questions) once exposed to a troublesome environment, will find it much harder than most to overcome it. The genetic effect has essentially been triggered, and appears to somehow relate to the brain's risk assessment function. The policy implication is clear. Early intervention when the signs of anti-social behaviour first emerge is crucial. Leave it too late, and turning it around is far more expensive and risky.

Of course many researchers not dealing with genetics already knew this, so again we can question the novelty of the information. For now, it is difficult to know what the practical implications would look like. Delinquency is hard to measure, and to a large extent culturally defined. (Risk takers are over-represented in prison yes, but they're also over-represented on the rich-lists.)

Trained teachers of new entrants can already identify the sorts of behavioural warning signs in five-year-olds that cry out for intensive intervention before it's too late, and they don't need DNA testing kits to do it. Yet still our society prefers to wait until kids grow old enough for prison, and spend our money there. It may be an attitude rather than a technological deficiency we should be facing up to. The risk of moving to a human model where we say to a person, 'You have the sort of genome that makes you more likely to be criminal,' is hardly negligible. The stories we tell ourselves about free will and personal responsibility are going to have to be very carefully constructed.

That's enough of a look inside the debate. As I say, it's a fiery wee meeting place, and those who like a good scrap will enjoy following up on some of the references. For me though, the really interesting question is what does this tell us about science and storytelling? Quite a lot, I think. For at heart, the Evolutionary Hypothesis manages to sneak a whole lot of story into its science, without people always noticing. Having done that, dishonest practitioners then have the chance to manipulate those stories in the name of science, in order to enhance not our shared understanding, but rather the status of their own prejudices. That is an excellent reason to keep a sharp eye out for such instances of story abuse. If we can clear the distractions from our eyes for just a moment, how much can science tell us about our nature?

The Evolutionary Hypothesis is important in so much as it reminds us that genetics play a vital role in our individual development path. However, contemporary studies of human behaviour provide the predictive (scientific) models. To this end, we should not be too greatly impressed by 'obesity the result of prehistoric metabolism says professor' type reports. Said professor is actually saying, 'There's too much food around, and we're finding it hard to say no,' and I'm picking we already knew that. The interesting obesity studies are those looking at how the modern metabolism works. What is the nature of the feedback mechanisms? How does the body regulate its own hunger and processing cycles? What is the role of insulin? How does diet affect gene expression within the pancreas? Tales from the savannah are important only if they lead these researchers to ask questions they would not have otherwise thought of. (Which remains an open possibility. Sparking interesting questions is a key job of any scientific model.)

The big abuse we most need to be aware of in this area is the 'women less interested in career advancement because of co-operative nature of berry picking' type hypothesis. This is nothing more than an attempt to give status to a contemporary prejudice by wrapping it in pseudo-scientific talk.

A final lesson to be taken from all of this is the danger of lapsing into fatalism, no matter what your viewpoint. Yes, you are what you are because of your genes, and the way they have interacted with your environment during your lifetime. Which of those two is the more important is not a particularly meaningful question. The key thing is that you and I have been gifted, thanks to the plasticity of our design, with the ability to do some pretty darned cool things if we want to. I can never win the Tour de France, and the reason is primarily genetic. No matter how motivated, nor how prepared to cheat I am, I still don't have the heart and lungs for it. However, to take the next step, and assume I could not, if I so desired, jump on my bike and cycle for four or five hours, because I'm not a born cyclist, is both limiting and ridiculous.

This is the problem with identifying average profiles. Saying, for instance, boys are physically stronger in the upper body than girls may be technically true, but it can also be tremendously unhelpful. The strongest women I have met could happily outlift ninety-five per cent of men, and most people, men and women alike, have little difficulty doubling their lifting capacity with a bit of training. So it becomes very sad if the message that girls aren't strong is the very thing that keeps young women from exploring interesting lifestyle options. So too with intellectual capacity. The range of our potential capabilities is significant, and life being short, most of our potential lies underdeveloped. The change in gender roles and expectations in our society in the last thirty years stands as the shining example of how much can be achieved when we are prepared to look past averages and treat people as individuals.

School teachers understand that the first thing they must do is provide motivation. From there, it's remarkable what the most apparently unpromising athlete, public speaker, actor or designer can achieve. I was a deeply average writer until I put in the practice. The first five novels I wrote were so bad as to be forever unpublishable. I'm a better writer now, I hope. Which of those two mes did my genes and my background create? Young people have the tendency to avoid the big challenges, having decided in advance 'I'm no good'. In that respect, it seems something close to criminal to spend too much time telling young boys that it is natural for them to be reluctant readers, or young girls that the sporting life isn't really for them. The less classified a young student feels by gender, race, family background, income level, body shape or whatever, the more likely they are to have a crack at being a human being, and this is surely a prize worth aiming at. In this respect some of the theorists in this field could be a lot more careful about the way they make their case.

We live our lives as if we have choices, as if it is within our power to affect outcomes. Free will, along with genes and environment, is the third component of our development path. We are the result of the decisions we have made. If we ignore this aspect of our humanity, then it doesn't matter which side of the argument we are on, we are still being depressingly deterministic.

I said earlier that the way our cells develop depends upon the feedback they are getting from their surrounding environment; this is nowhere truer than within the brain. The physical structure of the mind can be modelled as a huge network of connections. The strength and placement of these connections — in some yet to be understood way — underpins this activity we call thinking. These connections are not fixed things, but change throughout our lives, in response to our unique path through it. We are then physically altered by every thought we have, every sensation we enjoy, every experience we partake in and every decision we make. We are in the end not just the products of the development process, but also participants in it.

According to that view, there's an awful lot more to a human being than the genes they receive from their parents. Which is a problem when it comes to defining where a particular group of people came from… Time to visit a tropical island.

6 Sailing Away

'Reason is, and ought to be, the slave to the passions.' So said Hume and the point he was making is by now a familiar one. We can't reason our way to motivation: that comes from somewhere else altogether. However, our passions have a dangerous habit of subverting our reasoning faculties. When an idea packs a sufficient emotional punch we become reluctant scrutineers. One of my favourite examples, and the focus of this chapter, is the pleasure we take in tracing our ancestry. Mathematically misguided though this practice may be, it is usually harmless enough: no more than the diverting hobby of uncovering a direct bloodline link to somebody ancient, dead and famous. But when it comes to spinning the histories of entire races, it all gets murky and political, and we're probably obliged to make more of an effort to get things right.

Let's get the easy problem out of the way first. We have an odd tendency to get our thinking about ancestry absolutely back to front. Consider the family trees which the internet has made newly faddish. For good practical reasons, the aim is to go deep rather than broad. The mother lode — the fact will be paraded about the neighbourhood — is documented evidence that some father of a father of a father of a father ... cut William the Conqueror's hair. Starting from this illustrious forebear a chart is drawn, fanning down and out through time, all the way to you. There's something so romantic and alluring about this chain of serendipitous inheritance: a roll call of ancients battling, farming and trading their way across the world, taking risks, winning hearts, all leading to you. And William the Conqueror! His hairdresser. Hell, you're practically royalty. Who wouldn't be excited?

Well, a maths geek, for one. Rather than thinking of a family tree starting at some illustrious point in the past and fanning outwards, as genealogy can encourage us to do, we might more accurately think of the situation in reverse. For you had two parents, and each of them had two parents, as did each of those and so forth. Go back twenty generations (a mere five hundred years) and it looks like you need a million different ancestors! In fact we get by on far less, thanks to the fact that the paths keep recrossing, but the deflating implication is nevertheless clear. The only reason you're descended from royalty is that you're descended from so very, very many people. And if a particular king from centuries past has made it into your bloodline, it's an extremely good bet he's also made it into the bloodline of the neighbours to whom you so proudly insist on showing your scrapbook (unless they herald from a different part of the world). So where did you come from? The same messy mix we all come from, pretty much. Such a conclusion lacks the emotional heft we're after; so we ignore it. What is true for the amateur sleuth is also true, disappointingly, for the professional scientist. The lure of the big 'Where I Come From' story is sometimes just too strong.

As part of my research year, I looked at a very simple example of DNA extraction and analysis. I am no technician, but with help from an expert I was able to take DNA samples from myself and two others, extract mitochondrial DNA, sequence it and then compare it against a data base to draw some fairly dodgy pop-conclusions about the origin of my ancestors. I want to take you briefly through that process, and then show how a similar technique is being used to aid our understanding of human migration in the Pacific. In doing so, I want to again show how popular sentiment can draw indefensible conclusions from models which are essentially sound.

First, a little more of the technical stuff. Think of it as intellectual roughage: what it lacks in taste it makes up for in goodness.

Remember how a cell needs to master four tricks if it is to build a multicellular life form? Mitochondria are subunits found within the cell which specialise in one of those tricks: energy transfer. The thing that makes mitochondria so interesting to researchers tracing lineages is that the mitochondria contains its own DNA, which is to say part of the recipe for cell construction is held separately within the mitochondria. What's more, mitochondrial DNA is passed on exclusively (or at the very least, almost exclusively, there is some debate on this point) from mother to child. Your mitochondrial DNA came to you direct from your mother, who got it from her mother, who got it from her mother and so on. This is in stark contrast to almost all of the rest of your DNA, where you have two copies of each gene, one from your mother, and one from your father. These in turn are random selections from the two copies *they* each held, one from their mother and one from their father, so that the genes in your body are a result of a continuous jumbling and recombining down through the generations. (The other exception is the Y chromosome. If you are male, you have one of these, and you got it from your father, who got it from his father, etc.)

So the Y chromosome and the mitochondrial DNA represent known lineages. We know exactly the transmission path which led to them being in your body. The British researcher Bryan Sykes popularised this concept in a book called *The Seven Daughters of Eve*. The reasoning he used went something like this. Over time, mutations build up in our DNA, particularly in those stretches not involved directly in the building process. (Much of our genome can be characterised like this. Masses of noncoding DNA data punctuates the functional stretches.) Mutations might be thought of as mistakes; when a copy is made, a particular letter may be misread, although this in itself is a metaphor, as it assumes that making an accurate copy was the mechanism's greater purpose. The key thing about such mutations is that if they occur in the DNA which is to be passed on to the next generation, then the copying mistake is preserved. So if a particular mutation occurred in the mitochondrial DNA of some women thousands of years ago, then that changed sequence will be present in her children, and her daughters' children, and their daughters' children, and so on. From this, we might expect to see various patterns within the mitochondrial DNA of today's population, representative of the mutations accumulated along their matriarchal lineages.

Bryan Sykes figured that if he tested modern European populations, he might be able to put these mutation patterns into distinctive groups, with these representing their particular 'mitochondrial Eve', the earliest women to hold this particular pattern. He then suggested that the people of Europe fell, roughly speaking, into seven groups, each descended from a different 'Eve'. Each of these groups could be further characterised by particular geographical histories. So one group for instance might be particularly prevalent in a part of Spain, and in groups of people known to have migrated out of this area.

In the grand tradition of popular science, he sexed the story up a little, providing in his book speculative sketches of the lives of these ancient female ancestors, who by estimates lived tens of thousands of years ago, in the time before written languages, villages or agriculture. He then offered a service by which you could have your own mitochondrial DNA tested, and your lineage traced. This has proven as popular as, well, horoscopes is a comparison that springs to mind. Many thousands have paid their money and had the analysis done. And why not? It's a whimsical little story, a tale to tell the grandchildren, silly but diverting. It was exactly this process I mimicked in the lab, and in time I found that some mother of a mother of a mother of … my mother, may or may not have lived in Spain. Grand.

Or so I thought. But as soon as I told people about this experiment I was struck by two things: the intense level of interest, and the wilful disregard of any of the pertinent facts. 'Spain, really? Wow, that's fascinating. It makes sense actually, because I always thought that our family's dark features didn't quite fit the English profile.' Of all the stories we can tell ourselves, none is more powerful than the myth of origin: the tale that tells us we belong to some tightly defined group, which came from some particular place; that we have characteristics that set us apart; that our history is continuous and homogenous. Give people even the smallest scrap of identity data, and the instinct is to construct the most elaborate tales upon its flimsy frame. The 'this is who I am, this explains me' story is so incredibly powerful that we must be at our most very vigilant when considering the science behind its variants. Compare then, what mitochondrial DNA groupings can tell us, with what we would like them to tell us.

First, there are a number of technical issues which are conveniently passed over. The patterns we see are not as clear as the report back might suggest, so there's a certain amount of convention and educated guesswork going on when it comes to building family trees. Seven is a suspiciously convenient number: the sort that looks good on book covers. Sequencing is expensive and time consuming, so it has been common to choose small stretches of the mitochondrial genome for comparison purposes, and longer stretches do not always give the same results.

Putting this to one side, though — for as techniques improve these problems will lessen — there is a far more important point to consider. A fictional example may help here. I'm going to call her Amanda. She lives in New Zealand, with her part Māori mother and her father, who, as far as he knows, is of Irish stock. Amanda's mother's mother is a European, English originally, although a third generation kiwi, brought up in Taihape. Amanda's mother's father is the Māori connection, Ngāti Porou. On Amanda's father's side, his mother was indeed Irish, and came over to New Zealand when she was sixteen, although the man she believes to be her father is not, but that's another story. Amanda's paternal grandfather, despite also claiming Irish ancestry, was in fact adopted by an Irish family, while his genetic links are to Poland. This is the picture that emerges if we go back only three generations. We have genetic streams flowing from Māori, English, Irish, and Polish ancestry into the pool known as Amanda, along with an unknown wildcard. Yet, trace her mitochondrial DNA and we follow a single stream to Amanda's English grandmother. Even at a very simple third generation level, the mitochondrial connection is more remarkable for what it leaves out of the story than what it contributes. The tale your mitochondrial DNA tells may be a tale of dominant ancestry, or it may be the most insignificant of your genetic feeder streams. The woman to whom your mitochondrial DNA is traced may have contributed none of your other genes. Nor may the bulk of your ancestry have come from the surrounding areas and populations. Or it may have. The data is insufficiently rich to tell us either way.

There are two other points we have to make when tracing where our genes came from. You might remember that if you compare two human genomes, you will find far more similarity than difference. Most people carry the same versions of most genes. Otherwise we wouldn't be people. That being the case, whether you got a particular gene (coding, say, for an oxygen transporting mechanism in the blood) from a man in Italy or a woman in Mozambique, is in a material sense irrelevant if either way the copy turns out to be identical. It's a little like tracing the history of one of the water molecules in your swimming pool. Knowing it came from a particular rain drop, that was carried in a certain cloud, that condensed in a certain weather pattern, over a certain ocean, following a certain Christmas celebration on a nearby island, tells us nothing about the nature of the molecule itself, which is made — surprise, surprise — of one oxygen and a couple of hydrogen atoms. The gene has become part of modern mythology, and in the popular world at least, we are starting to grant it mystical qualities, because that is what we have always done when discussing lineages.

You might also remember that when trying to pinpoint the start of the scientific revolution, history doesn't provide its own punctuation. Any point on a journey you choose as a marker is arbitrary. Where did your ancestors come from? Australia? Africa? The jungle? The ocean? How about The Big Bang? Creation as we know it is a continuous process. Sure, a particular hyper-great gran may have kicked about in Spain, but where did her mother come from? And her mother? We tend to choose the most romantic starting points, those that most capture our imagination, and also those that are the most accessible. This is for artificial, self-serving storytelling needs rather than modelling purposes.

Which brings me directly to a body of research which has been flavour of the month with the popular media. The question researchers are asking is, where did the people of Polynesia come from? Already we should be alert to the dangers behind this question. The answer depends on how far you want to go back, presumably, and it's always going to be 'many places'. In part too, the answer must be, 'nowhere, for they are the people of the Pacific'. And a sensible answer will always include the question 'which particular person are you talking about?' Even the more abstract 'it depends what you mean by a person' should get an airing, if we want to address this question with any sort of rigour. But I mentioned the media, so perhaps rigour is not going to play a major role in this story. Rather, I fear, the answer they are going to dish up is 'Taiwan'. That, I hope to convince you, is an inadequate answer.

While writing this, I came across two major newspaper articles, two features in national magazines and one television documentary on the Taiwan hypothesis. Two further television documentaries are in the offing, and bizarrely, one high profile politician has become fond of trotting it out as evidence that he, personally, has nothing against Asians.

Let's have a look at the evidence first, to see what all the fuss is about. From there, I'll examine the way the media are interpreting it, and make some guesses as to why (I'm a story teller too).

It is undoubtedly true that people at some stage settled the many islands of the Pacific. Looking at that part of the globe on my desk — an entire hemisphere of ocean speckled so very lightly with islands — I can't help but be impressed by this fact. Somehow, people found all this stuff. That suggests some fairly wild ocean adventuring, a tale that immediately plays to our romantic tendencies. The question that interests many people is: 'who arrived where when, and where did they come from?'

What people seem to be looking for is a great big flow diagram, with neat simple arrows denoting boatloads of people packing up and moving on to the next island. Although such representations are a simplification of what probably took place, with multiple settlements, return journeys, the maintenance of trading links etc., the Pacific story lends itself to this sort of analysis more than most. The fact that the area is mostly water constrains the amount of mixing between peoples that could occur, and with all the evidence currently suggesting that up until a few thousand years ago, none of these islands had people on them, it does make at least some sort of sense to talk about broad patterns of movement.

How can we recover these tales from the past? Scientists rely heavily on three different sources of evidence: physical artefacts, language patterns and DNA sequences. People leave many physical clues long after they have passed, be they shards of pottery, seashell fishhooks or the remnants of ancient fireplaces. Digging about can reveal all sorts of hints regarding the people who once inhabited a land. Modern carbon dating techniques help. Of course, the leap from a few artefacts to pictures of ancient cultures is a big one, and some of the stories woven together to make sense of the discoveries are just that: fanciful works of imagination. We are dying to hear something more than 'We found this particular piece of rock, and it looks like it might have been used as a tool of some sort, but really, who the hell knows for sure.' Far more exciting to speculate 'It was used to strip the skin from a downed wildebeest after an exhausting pursuit by the tribe's three most skilful hunters, in which a young boy had his thigh bone shattered, causing him to later die an excruciating death from blood poisoning.' We would be foolish to think archaeologists haven't upon occasion succumbed to the storytelling temptation. Nevertheless, the more the evidence accumulates, the more constrained these stories become.

Linguistic evidence is another smart technique for piecing together the past. Language is constantly changing. Terms come and go, and new words are needed to cover new circumstances and concepts.

When people become geographically isolated, their languages slowly move away from the language of their place of origin (and that host language will in turn be changing). It is not unreasonable therefore to suggest that the level of similarity and difference between languages has something to tell us about how long people have been apart. Linguists construct language families based upon just these patterns, and make suggestions as to which languages are related, and how. Again, in a model of relative isolation like the Pacific, this can add a lot to the migratory jigsaw.

Where populations are more continuously connected however, with a lot of flow backwards and forwards, it becomes much easier to think of ways that the flow of language and the physical flow of people can become separated. The language of a dominant political power or religious tradition can spread with influence rather than with genes. Consider the impact of the Roman Empire on many modern language traditions. Equally, trade is made easier by people learning common languages, and so language can spread on the back of commerce more quickly than it is ever transmitted down the generations, the popularity of English today providing a good example. So while there is definitely a signal contained in linguistic information, again it is a scrambled signal, more applicable to some situations than others. Coherent accounts are stitched together with the threads of story.

DNA analysis shares many similarities with its linguistic counterpart. As discussed with the mitochondrial example, people take ancestral signals with them in their genomes, and over time these signals are added to by local mutations. The mutation rate for DNA is far slower than that of language, and so broader, older patterns may emerge. Equally, as we have seen, the way DNA recombines as it is passed down through the generations, muddying the signal, limits the usefulness of data. But small populations moving into new areas will often take with them unique signals which will be preserved through time. Researchers have not just restricted themselves to the DNA of people. Living things which are carried with moving populations, crops and animals for instance, can also yield important clues.

Currently, analysis of kumara and gourd DNA is being used to try to add to our understanding of what sort of contact might have occurred between the people of the Pacific and South America.

This type of research is carried out by experts who by and large are trained in the scientific virtue of caution, although the vagaries of publication and status demands mean we need to be wary of compromises made along the way. So where is the picture they are building up tending towards at the moment? At some stage, about 1000 B.C., people appear to have moved out from Near Oceania, specifically the Bismarck Archipelago north of Papua New Guinea, into the Pacific. Popularly known as the Lapita people, they are named after the first discovery site of the distinctive dentate pottery they are associated with. (It is fun to wonder what lasting artefacts our own culture may one day be remembered for. Will we in the future be known as the Concrete Block people?) Over a fairly compressed time frame these people appear to have settled New Caledonia, Samoa, Tonga and Fiji.

Over the next two thousand years, people spread out through the rest of Far Oceania, discovering the far flung lands of Easter Island, Hawaii, New Zealand and all else in between. By piecing together clues about prevailing winds, and the sailing technologies of the time, and adding information from oral histories and the data already discussed, scientists can make good guesses regarding the sequence of this exploration. The tale is a rollicking good one, bookended by the initial movement of the Lapita culture down from the Bismarck, and by the world recorded by the first Europeans to arrive in the area, thousands of years too late to be called explorers. This is sure to be a massive oversimplification of course, although it is nothing compared to the cobbling together of the stories of earlier origins.

Remember the amazing pull of the 'where do we come from?' question. If we say the people of Polynesia came out of the Lapita construct, we just know there is going to be another question. Yes, but where did those people come from? Here we get into the murk of definitions. What does it mean for a people to come from anywhere?

People have been in the Bismarck region for some thirty thousand years. Papua New Guinea and Australia were settled even earlier, so we're getting right back to the dawn of the modern human here. In fact the Aboriginal Australians deserve the title of the first great ocean voyagers, as at some stage in the distant past they had to cross water to get to Australia. The genetic, linguistic and archaeological evidence all point to a far more recent arrival of people in the Bismarck Archipelago: only some five hundred years before the expansion out into the Pacific. Coming out of Asia, these people brought new language, technology and genes into the region.

A simplified model might suggest it was this people who arrived, built their stilt houses over the lagoons, hung about a while, intermingled with the locals, developed some new sailing technologies, and then trundled off to explore the vast expanses of the Pacific. And if you summarise it quickly enough, you can almost see it as one group of people: as if they just stopped off on their way for a chat and a cup of coffee. But you can drink a lot of coffee in five hundred years, and do a good few other things besides. If we are to make the tempting mistake of tracing the people moving into the Pacific directly back to those arriving from Asia we can miss some questions that to me, anyway, feel important.

In what way exactly were the people who arrived in the region similar to those who spearheaded the later exploration of the Pacific? What does it mean to call them the same people? Yes, genetic, linguistic, cultural and technological traditions flowed through them, but in each of these respects there must have been interaction with the locals and their environment. So why do we say Austronesians moved into the Pacific? What is it that encourages us to accentuate the contribution of the latecomers at the expense of those who had always been there? We can talk about the degree of intermingling, and use genetic and linguistic profiling to model these, but is there any objective measure by which we can say that a people has maintained enough of its identity to be considered a hermetic cultural unit? If there isn't, what do we mean exactly when we tell these stories? We use the tools of science to add to the tale sure, but really, how scientific are we being?

Consider for illustration what we would make of the following fantasy. A group of young New Zealanders on a *Spirit of Adventure* sailing trip are blown off course during a cyclone. People assume the ship has gone down in the storm, but in fact the youngsters are left stranded on an obscure and rarely visited island. This group of thirty enterprising young New Zealanders make the best of a bad situation and in time become the founding population of a whole new community. Of course, conditions are harsh, and not all survive. Those who go on to reproduce are only a subset of those originally on the boat. For reasons which I can not be bothered inventing, it is another eighty years before anyone visits the island. By then all the original inhabitants are dead, along with their children, and their grandchildren and great grandchildren now manage the small community. So where did they come from? The people are asked, but know only that they came on a great ship. The language, artefacts and frequent references to Jonah in their myths and legends all suggest New Zealand might be a good answer.

But what say we use DNA techniques to confirm this hypothesis and find that the markers suggest a mixture of Asian and European origins? It's quite possible. People of Asian and European stock were on board, and one chap in the founding population was particularly fertile, skewing the sample. Other influences, Māori, Pacific, and a lad from South America, are not found. How reasonable is it to say the island has been populated by the people of Asia and Europe? To my thinking, it's problematic. They were New Zealanders, made into the people they were by the influence of that country as they experienced it before being blown off course. They brought with them tales of rugby, summers at the beach, backyard hangis, Pacific beats and Auckland traffic jams. The link to the more ancient ancestors of some of the island's founding population seems to me to subtract information, by distracting us from important cultural influences. You can see to what extent subjective judgements must cloud our discussion.

The issue of origin and identity is a troublesome one, and I am not here proposing any fixed rule for resolving it. Rather I am pointing out there is no fixed rule, something we need to bear in mind when we fill in our maps with their little people-movement arrows.

Again, science can provide us with the data, but not the interpretation. It's worth wondering what, in 2006, are the cultural biases which might promote the importance of an Asian source over one a little closer to home. Do we also have a story bias which encourages us to see cultures as units which travel intact through time, rather than as poorly defined, amorphous masses of interaction? It would be no surprise if the appeal of certain interpretations can be traced back to our most comfortable prejudices.

But what about Taiwan? Well, this is just the same complication writ large. Linguistic and genetic clues both lead us to Taiwan as a possible source for the proto-Austronesian culture, although you are right to wonder what this statement actually means. The Austronesian family of languages, of which the languages of the Pacific belong to one branch, seems to have its source amongst the indigenous Taiwanese population, and some mitochondrial motifs found amongst this population also occur in the Pacific. How romantic is that? And how cool does it make modern scientists look? We have traced the story back five thousand years, all the way to the birthplace of the people of the Pacific. Clever old us. This is where the media get really excited. They can produce impressive graphics, formidably accurate date estimates, and animated models of migration. They can send reporters to Taiwan, where they will be filmed being astounded by the clear 'Pacific vibe' of the locals.

Amazing stuff. But, I hardly need tell you, simplified close to the point of meaninglessness. At this stage, do we know the mechanism through which the language spread down through parts of Asia? To what extent was it via displacement of people, and to what extent just displacement of linguistic tradition? Even if such human displacement occurred, to what extent did interaction between the peoples drive the changing cultural and genetic mixes? Does the transmission of mitochondrial signals reflect the broader patterns of breeding, or are some interesting coincidences luring us into a false interpretation? Even as supporting Y chromosome data emerges, we must ask what the practical limits of our knowledge look like. It is a complex situation to model, and the data we have is puny by comparison, so the grand traditions of narrative again take over. Taiwan just seems like such an unlikely and exotic starting point, and we are drawn to it even though as a genuine source it has similarities with our *Spirit of Adventure* conjecture. The sceptic understands that no matter what mythical homeland we choose, emotional reporters will quickly find the sort of startling cues of similarity upon which astrology is built. (Look, they walk on two legs, just like we do. Our long lost brothers!)

People appear to prefer the simple story — the one that says we are especially related to some groups and less especially to others — over the more mixed up but just as valid alternative narrative: that we're all a little bit of everything, one way and another. We love the idea of identifiable 'peoples' travelling through time and space. It helps us get a sense of who we are. And of course such a model has important political ramifications, which guide our interpretations. Far be it from me to criticise our preferences. Stories that appeal are, at the end of the day, stories that appeal. But we must acknowledge that that's what they are: stories.

It is also good to be aware of what might be called The Governor Effect. Back in the seventies, New Zealand television made an historical drama series called *The Governor*. A lot of money went into it, and great effort was made to reflect the historical scholarship of the time. A generation were raised on this powerful vision of our past. It became our story. Of course scholarship then moved on. Historians uncovered new stories, and promoted new interpretations. But it is not sensible in a world of limited resources and attention spans, to produce and show an updated drama, to put the record straight. The media snapshot became the received version.

I can't help but think the Taiwan hypothesis is about to set in the public consciousness in much the same way. 'Where did the people of the Pacific come from?' 'Taiwan.' Is that what our children will be taught? Who is to blame for this? Maybe no one needs to be blamed. Maybe there's no problem here. Could it be that simplified stories are all we need? To me that somehow seems to diminish the importance of so much careful scholarship over which the blanket of story has been thrown. I think we should blame ourselves, and our appetite for simple, politically palatable tales. When it comes to constructing stories, we tend to find what we are looking for.

Loitering about at the edge of this whole issue is an even more radical challenge to the construction of origin tales, and I will finish with that. During 2005 a small controversy flared up when some Pacific academics spoke out against a large scale DNA sampling programme designed to flesh out the tale of human interrelatedness. The argument went something like this: 'We don't want your Western techniques coming to tell us our stories of origin. We already have tales of where we came from. This is yet another version of colonisation.'

At least one local academic responded sharply in the national press. His argument was that this isn't storytelling, it's science. And to resist such science is to promote ignorance. This, he suggested, was just political correctness gone mad. Notwithstanding the fact that I still don't know what political correctness means ('I don't like your views, they make me angry' perhaps?) I do think the attack is misguided. For science, as we know so well, has a habit of quietly slipping into storytelling mode when nobody is watching. There is something distasteful about telling stories about others, and then using the force of our institutions to impose these stories upon them.

During my research year I had the privilege of travelling to Samoa for a 'look around'. I lay no claim to being an expert on Samoa, indeed upon landing my ignorance was complete, and leaving a fortnight later I felt only marginally better equipped to pass comment. Nevertheless the obvious outsider's observations can be made. It's hot there. (That sounds like a typically inane tourist's observation, but bear with me, I'm working my way towards a point.) Climate surely has an impact on the sort of people we become. To take just one example, consider what happens when you combine warm days and muggy nights with the available timber. The Samoan fale is a fascinating piece of architecture: perimeter poles hold up a thatched roof, and woven walls are lowered only when the elements demand. The result is a type of open plan living which most westerners would, I suspect, find rather challenging. No interior walls to impede air flow, and for much of the time a clear view of everybody in your own home, and in the homes of all your surrounding neighbours.

I spent a lot of time speculating what this sort of arrangement might do to a community's values and perspective. I don't know which of my fantastical answers is most correct, and I'll leave you to your own imaginings, but surely such an architectural detail has the potential to profoundly influence social interactions, and hence sense of self.

Consider also the sorts of foods available in a tropical environment, from fish caught in benign fishing conditions, to abundant fruit. How does this affect the very notion of being a human being, when compared to the effects of a climate where every last calorie must be clawed from a reluctant land? Many other unique factors can be woven into the Samoan story: the two main islands within easy distance of one another; the rugged mountainous interior; the compact size of each land; the relative isolation maintained over thousands of years; the sorts of epidemics that flourish in such a population and environment; the unique stories that weave their way through the land's long history; the consistency of the length of day and the lack of extreme seasons; susceptibility to tropical storms; the link between the language that has developed and the patterns of thinking associated with it — and so it goes on. Samoa is truly like no other place on earth, and fa'a Samoa, the Samoan way of life, reflects this. If you ask the locals where Samoans came from, the popular response seems to be, 'We come from here. We're Samoan.'

If the critical academic mentioned earlier was correct in his attacks, this is a naïve answer, ignoring as it does the good evidence that

Samoa was peopled from the west some three thousand years ago. But isn't the scientific answer also in danger of lapsing into naïvety? The people who travelled to Samoa all that time ago may have been many things, but they were not Samoans. To be Samoan is to be part of the unique cultural construct made up of a physical, cultural and genetic environment. We know you and I are not constructed only of our genes. We are the sights we have seen, the food we have munched, the thoughts we have entertained, the stories we have heard, the land we walk. I don't mean this metaphorically; I mean it in a boring, reductionist sense. Were it possible to take a snapshot of ourselves in time, detailing the structure of every last cell, every synapse, every memory network, every predisposition, then the physical manifestation of all these apparently abstract concepts could indeed be found. That this is beyond science, and will be for the foreseeable future, doesn't make it any less scientific.

It is a failure of imagination to believe that only the things we can measure have any relevance. 'Not everything that is measurable is important,' said Einstein, 'and not everything that is important is measurable.' It is the most natural thing in the world to think of humanity in terms of a continuous flow (just as it is natural to think of the individual person flowing through time as a single entity), each new human being growing out of the contribution of their parents. But natural metaphors need not be accurate metaphors. You and I are a mixture of many things: genetic material, physical environments and cultural influences. Every thought we have had, every story we have been told, has literally, physically shaped the connections of our brain, in the way that every mile we have walked has had its impact on our muscles and our cardiovascular system.

Consider a hermetically sealed culture which is one day visited by a stranger who teaches the locals to read and write, and leaves them with a functioning, high-speed internet connection. The stranger then departs, leaving not a biological trace. Yet a hundred years later, that society may have changed dramatically. The people are not the people they would otherwise have been; the changes are dramatic, and they are physical. To say that people can be modelled purely in terms of a physical stream, that the visitor is not an important part of the 'where did they come from' story, is also naïve.

I am not arguing that there is no validity in using DNA to add to the picture of human migration. But it is only a small part of a larger story, and we must be aware of the cultural influences which encourage us to latch on to this particular aspect of the tale. (It's scientific. It must be right!) I come out of a European tradition where the obsession with lineage and origins owes a debt to a history littered with invasions, migrations and cohabitations. But it pays to be careful when assuming that the same sort of approach to the explanation of origins should hold sway in the collective mindset of a place with a history like Samoa's. When next someone tells you where a particular people come from, you might like to ask them what they mean by a people.

We must also be aware that the more power we have, the greater our access to respected journals and influential media, the greater our responsibility to ensure that we do not hijack other people's stories. The right I have to tell stories about my own existence is, to my mind, fundamental. I would not welcome a scientist with a clipboard coming into my house to analyse my lifestyle and beliefs and disabuse me of my most comfortable myths. Especially if all they wished to replace them with was the anaemic sort of reductionism that genetic analysis at its barest represents. We have every right to make available our models to the world for due consideration, but when this slips into imposing our own form of mysticism onto others, I get mighty nervous.

To finish on a cute little story: five hundred kilometres north of Samoa is the atoll group of Tokelau. The people there have their own rich set of stories regarding their origins. My brother, who spent a year living on one of the atolls, asked one local where he thought his people originally came from. 'Atlantis,' came the tongue in cheek reply. 'It sank and we came here.'

'Really?'

‘Well you know, it’s just a story, but it’s as good as any other. It suits me.’

He has a point, I think. As good sceptics we must accept that truth is a mystical and unhelpful concept. Its replacement — usefulness — demands a new sort of intellectual discipline. We tell ourselves the stories which best serve our purposes. Does anybody else have the right to tell us what that purpose should be? That brings me to one of the most powerful narratives of all: the story regarding the mind, and what it means to be human.

7 Blue Sky Dreaming

'Why's the sky blue?' a child asks.

You run through the usual responses.

'The same reason the grass is green, it just is.'

'Well, actually, son, it's all to do with wavelength and the refractive index.'

'Nobody knows, for sure.'

Oddly, all those answers are right, in their own way. It depends how you choose to look at the question.

This chapter isn't about the sky though; it's about the mind. What does it mean to think? Are animals conscious? Could a machine ever be? Do I have free will? How does that work? What is the relationship between the things the brain does, and the way that thing feels to me? I won't pretend I'm about to offer a definitive answer to any of these, for as with the sky example, it all comes down to what sort of answer you're prepared to accept as adequate. Instead I want to look at the twisted philosophical arguments that questions of consciousness have given rise to, and whether they have something profound to say about what it means to be human, or whether it's really just a party game for pointy heads. Of course I will be trying to see if the sceptic's viewpoint has anything to offer: whether the distinction between science and storytelling might not be an excellent way of prising the combatants apart.

First, though, let's return to the question of the sky's colour, which oddly enough captures much of the flavour of the consciousness debate. Why is the sky blue? There are a few things we need to know if we wish to approach a vaguely scientific explanation. The first is that colour is a function of light's wavelength. Different wavelengths excite different combinations of reactions in the receptors in our eyes, which in turn send different electronic signals to our brain, which match different colour perceptions. Echoing Kant, we might say colour is less a quality of the world, and more a quality we assign to the world. Along with visible light, there is a range of wavelengths our eyes don't respond to, including things like X-rays and ultraviolet light. The light zipping down to us from the sun, white light, is made up of the whole spectrum of visible light, which is to say all the different colours with their various wavelengths, and of a range of invisible (to us) rays, like the UV we're forever being warned about. We can see the colour spectrum contained in sunlight when water droplets in the air split the light reaching our eyes into its constituent parts, leading to the appearance of a rainbow.

We need a way of thinking about the relationship between wavelength and the behaviour of light, and I'm going to settle on an arresting, albeit unscientific, metaphor. Short wavelength light is more likely to be bumped off course than light with longer wavelengths. When I explain it to younger students, I like to describe short wavelength light as being particularly scatty, with a short attention span. It's the ADHD child of the light world. It sets off from the sun with good intentions, meaning to travel directly to earth, but as it hits our atmosphere and encounters the many tiny molecules contained within it, it gets, well, distracted. At each interruption, it is liable to zip off in a new direction. So short wavelength light doesn't travel directly to us from the sun, like well-behaved light might, but scatters itself about the whole sky on the way. The light we perceive to be coming from the sky, the light that gives the sky its colour, is the scatty light, making its way back from its various distracted paths to our eyes. And because it is the blueish light that has the shorter wavelength and so is the most easily distracted, the sky appears blue.

The more sober, responsible and difficult to distract forms of visible light, the reds, oranges and yellows, arrive instead in a direct line from the sun, and the combined effect gives the sun a brilliant yellow/white appearance.

Like any good scientific explanation, this also unlocks other mysteries, and gives us some nice predictions. Here's an interesting one. Late in the evening, as the sun is setting, the angle between the sun and the observer is far shallower, which means that the light has to travel through a thicker slice of the atmosphere in order to reach us, which in turn means there are more potential distractions. So we might expect that around sunset for instance, there is a greater chance of some of the normally well-behaved red light being distracted across the sky. This is exactly what we see (magnified by the fact that once the sun has dipped below the horizon, its direct light can't drown out the distracted signals). We might even predict that this effect will be exaggerated by events that put more distracting material out into the atmosphere: city pollution for instance, or a volcanic eruption, and again we'd be exactly right. So there you have it, the colour of the sky explained, and all within the context of our simple model of light and wavelengths. What's more, we get rainbows and sunsets thrown in for the same price.

Excellent. Who wouldn't be satisfied, ready to now move on, and accept that we understand the blueness of the sky? Yet I claimed that it is also quite reasonable to say we don't know why the sky is blue. What am I getting at?

Well, the truth is that my neat and appealing explanation only works if I am allowed to pull some dirty tricks. Vital issues are skipped over as if they are not relevant. For example, why are short wavelengths more easily distracted? We can answer that one without too much trouble, but not the next. Why does light travel as a wave at all?

That's not a simple question, for in fact light appears to have a dual nature, behaving as both a wave and a particle, and that behavioural difference takes us straight into the tangled heart of quantum physics.

When I say it's tangled, it isn't that I don't want to get distracted by an explanation. Even if I did want to tell you how light manages to be both things simultaneously, I would have to choose between rival, and immensely unsatisfying suggestions. At the conceptual level, it's fair to say this is a problem still being worked on. If we push our questions far enough, we run into a wall of bluster, and if we manage to scale that, we find ourselves staring up at a mountain of ignorance. It was always thus. Am I saying, therefore, that we do not really understand why the sky is blue? Not at all. In everyday terminology we have a jolly impressive explanation. I'm just pointing out that at the technical level, explaining something simply means pushing the mystery back a little further, and for most of us, most of the time, that's more than enough.

Think about what happens when you take your car to a mechanic because the engine is sounding rough. Depending upon your knowledge and interest, there will be a level of explanation you find satisfying. ('Yeah, I've fixed it,' usually does me just fine.) But few of us will expect the explanation to go as far as the fundamentals of the physics involved. It's an efficiency thing, meaning we prefer to deal not with the world as it is, but rather with the metaphors which best serve our utilitarian desires.

This may seem like a silly and obvious point to be labouring, but very soon we are going to meet an argument over whether or not consciousness has been explained, or ever can be explained. Part of this is simply an argument about definitions. Our explanations of consciousness are getting better, just as our explanations of light are too. But there's still plenty of stuff we don't know, and there's probably stuff we'll never know — just as with all endeavours. The key question then, is can we ever know enough about consciousness to satisfy our everyday curiosity, or will the mystery element always be so in our face that technical explanations remain unsatisfying?

To get a small taste of the controversy, there is in the field of mind studies an even bigger problem I carefully avoided when I gave my blue sky explanation. Why does shortwave length light appear bluer to us, and longer wavelength light appear redder? Can we even be sure it's the same way round for other people? (They might see the sky as red, but call it blue, just as we do, for that is the language they have learnt to use.) We know the light, by dint of its wavelength, stimulates particular chemical reactions in the eye, which in turn send a particular pattern of impulses into the brain, which in turn cause particular patterns of brain cells (neurons) to fire, which in turn lead to the colour blue being represented — but what the hell does that last bit mean? 'Blue being represented.' Hmmm. Sounds suspiciously like bluffing. What is the link between certain bits of my brain lighting up, as it were, and the sensation I get when I see blue? Who is this 'I' I am talking about, who does the seeing? Is there a little theatre there, where the blue is projected up for me to look at? Well, if there is, then we're in trouble, because we know only how the image is projected, not how it is viewed. It would tell us only how the brain displayed blue, not how I took the message from the brain and 'saw' blue. And that can't be quite right, can it, for it rather begs the question, why have a brain at all? Why not have the little invisible guy called me inside my head just look up at the sky directly and cut out the middleman?

Perhaps the most interesting question of all, when we ask the question, 'Why is the sky blue?' is 'What does it mean to see blue?' How does this sensation actually occur?' In fact, what does it mean to perceive any quality: to feel wetness, to experience yearning, to succumb to curiosity, to hold arguments with yourself inside your head, to be battered by grief, to turn the words you are reading now into personal meaning? In other words, the whole business of being human. What's that all about? Even if we can show it is just the result of the boring chemical reactions within our brain, why do they play out the way they do? That's the big question. It's the question some philosophers and neuroscientists like to think of as the hard question, and which those on the other side of the great divide like to dismiss as a logical trap we have fallen into, as no real question at all.

I don't think you have to be a genius to work out that this debate is going to be a heated one. Should it ever fire the public imagination — and its esoteric nature means that probably it won't — it has the potential to be even more intense than the evolution debate. Why do I say this? Because it is messing with our most sacred metaphors, challenging us to re-examine them. And like moving house, this re-examination is an extremely stressful, high-cost process, one we don't undertake lightly.

Prior to Descartes (1596–1650), it had seemed natural to assume there was such a thing as the essence of life, some weird spiritual quality, breathed into all the animated creatures that scurried, slimed, wriggled, hopped, swam and flapped their way about our kingdom, and even breathed into all those rooted plants that miraculously grew from the soil. This was just one of the central mysteries of existence, a reason to treat the creator with awe and reverence. Perhaps though, Descartes challenged, what we see is an illusion. Perhaps the so-called living forms are nothing more than automata, clockwork expressions of the laws of physics, and it is our own imagination which breathes the life into them. It was another two hundred years before the evolutionists provided a model solid enough to back up the hunch, and now, for many of us, life itself doesn't seem quite so inscrutable. As happened with the colour of the sky, we have stripped it of another layer of mystery.

Descartes put a fence around one aspect of the human machine: that of consciousness. The same sort of reductionism that has over time yielded such powerful results when explaining the mechanics of growth, motion and reproduction, could not, he supposed, explain thought. Thought was some other category of existence altogether, linked to the workings of the physical brain, but in some mysterious sense, removed from it. This is what we think of today as Cartesian Dualism. This brand of thinking has one great advantage, and one great problem.

The advantage is clear. It lessens the psychological blow of collapsing life to simple mechanics. There is still an unknown and unknowable essence. The human soul, in short, is rescued from the tyranny of mechanics. Nor are we asked to question the nature of our most certain data, that of our own subjective experience. Dualism leaves some of our most precious and fragile ideas carefully undisturbed: nor are we asked to question the nature of our most certain data, that of our own subjective experience.

The disadvantage is equally obvious. If we propose two worlds which are both separate and intertwined, then even if we are happy to accept one is beyond our understanding, how are we to think of the way this world exerts its influence on the world we can model? How does thinking stuff change the way active mechanical stuff goes about its business? And shouldn't, by implication, this exchange be open to our investigation? Descartes himself was acutely aware of this dilemma, and rather than run away from it, he spent much of his later career wrestling with it. It doesn't appear he ever found a solution he was very satisfied with. The question is, have we?

I want to approach this chapter in two distinct ways. First, I want to look at some of the advances we have made in our understanding of the physical part of thinking since Descartes' time, and in particular over the last fifty years. We have discovered many weird and wonderful things about the brain and how thinking occurs, and some of them have clear implications for the way we view consciousness. Again, I'm not proposing a learned overview. I'm just going to present a grab bag of the bits that most fascinate me. And I will try my best to highlight the fact that the brain, in all its complexity, is still a tremendous mystery to us.

I want to then leap from the scientific to the philosophical. The question is a simple one. Is this the old life argument replaying itself? As we dig deeper into the brain and its workings, are we beginning to see that consciousness is largely an illusion, which is to say more of a metaphor, a way of thinking about relationships, than an actual phenomenon in need of scientific explanation? Or, instead, is consciousness more real than any variable we may wish to model? Is this the classic case of science having met its match? Could it be that the human brain just isn't wired in such a way that allows it to understand itself? Some of the last part of this chapter gets a little abstract, so it's a good part to read slowly. If you're feeling tired, put it down and come back to it in a fresher frame of mind; the ideas in here reward slow contemplation.

And so to the brain. There is an easy trap which I sometimes fall into, when I am being lazy, and slightly sentimental. 'Of course,' I find myself thinking, 'there's more to thought than this simple mechanical process inside our brains. I mean, look at the brain. It's just a sloppy hunk of meat, a one and a half kilo lump with the consistency of custard. You can't tell me you can get an appreciation of poetry out of that.' But such reasoning is as ridiculous as looking at a cellphone and saying, 'It's just a little hunk of plastic and metal. Of course I can't send a photo around the world using that.' Our crude senses are a poor tool for understanding complexity, particularly when we're talking about something as mind blowing as the human brain.

Maybe a number will help you to switch perspective. How about 10^{15}? Doesn't look too impressive just like that does it? But remember earlier I told you three billion was almost unimaginably large, more than the number of seconds in a human life? If you were to lay three billion one cm blocks end on end, they would take you right around the world, and then some. Okay, well it turns out three billion is a pretty small number. 10^{15} is a million lots of a billion. Now you have enough little one cm blocks to go around the world about 400,000 times! No, please, stop and think about that. Think about going outside and laying down the blocks. Think about the ridiculous number it would take just to get you to your nearest airport. But right around the world? 400,000 times! So an unbelievably, stupidly huge number. And yet, 10^{15} (a million billion) is about the number of synapses in the human brain.

That's going to be a little more impressive if I tell you what a synapse is. It's a connection point between brain cells, essentially. Axons, long tendril-like affairs, reach out from brain cell to brain cell, and at the point of meeting we have a synapse, a gap over which a chemical message can be passed, and converted into an electrical current which moves to a brain cell (neuron), so changing its internal state, and determining whether it will in turn pass the message on, and if so, with how much vigour, and along which of its many connections. Signals are buzzing about in our brain in this way at something like 400 km/h, and the number of potential transmission points, the synapses, is, as I say, a really, really big number. So big I can't think of a way of making it make sense to you (or me) in practical terms. I don't have the expertise to explain the workings of the brain in any great detail, and I'm not even going to try. But I do want to make the point that the vast complexity of the brain, the sheer potential for combinations of firings, is truly beyond our current understanding; and in my case, even beyond my conception. This is a very sophisticated piece of machinery, the most sophisticated piece of design in the known universe, no less. So, before we go dismissing it as no more than an item in a butcher's display window, incapable surely of generating the mystery of abstract thought, we would do well to be aware of just what it is we are dismissing.

Another point, which follows on, is the question of how on earth the 10^{15} (a million billion) connection points between the brain cells know where to put themselves. For, believe it or not, they represent only a small portion of the potential brain cell connections. This returns us to our discussion of two chapters ago. The simplistic view of genes as the whole construction blueprint needs modifying when it comes to the building of the human brain. Brain connections appear to be formed during development, and strengthened and altered over our subsequent lifetime, according to the use they get. To use a simple example: take the case of an Italian boy who grew up blind in one eye, despite the fact that the doctors could find nothing wrong with the physical make-up of the eye itself. His blindness was something of a mystery until it was discovered that the boy had worn a bandage over the slightly infected eye for a short time when he was very young. Unfortunately, this was just the time when the connections were forming between that eye and the part of the brain used for sight processing. During this phase, the brain was literally waiting for the signals from the outside world in order to provide it with the necessary connection information. And when no information came from that eye, the connections were not formed.

Another nice example can be seen with particular species of birds, which are programmed to come to think of the first creature they see as their mother. Once the connections are made, there's no telling them otherwise, and if they first clap eyes on a person, they will happily follow that human minder around in the manner of a devoted, trusting child. Think of it perhaps as a word processing programme, capable of setting out the finest novel or the most turgid piece of technical prose, but helpless to do either without input from the external world: the keystrokes. Our brain constantly responds to its environment. Our thoughts are, to recycle again, the brain's exercise programme, building up physical networks, ways of thinking, perceiving, responding and therefore being.

Before I go any further there is a point I have to slip in soon, lest the guilt overwhelm me. The trouble with discussing anything as sophisticated as the human brain is that you are forced to use simplified metaphors (10^{15}, a million billion, is just too big to think about on its own terms) and sometimes these metaphors can cause a lot of trouble. We slip naturally into talking about networks, information flows, processing, wiring, and parts of the brain specialising in certain tasks, yet none of these terms, many of them borrowed from the language of computing, is likely to be terribly accurate. The gap between our conceptual models of the brain, and our nuts and bolts understanding of the technicalities is large, and we use jargon as a sort of filler. It is important to remind yourself as often as possible that 'it's just a metaphor'. We can't rule out discoveries in the near future than could turn all these metaphors on their heads. (One intriguing possibility on the horizon is the theoretical development of quantum computing programmes.)

Bear that in mind as I now plunge headlong into some of the studies of brain damage that have changed the way we look at brain function over the last fifty years or so. Perhaps most well known of all these studies are those of patients who, having suffered epilepsy, were treated by a procedure that severs some of the connections between the two hemispheres of the brain. This limits the sort of communication that can take place between the hemispheres, and where the malfunction of this particular communication contributes to the illness, severance can alleviate the symptoms.

Yet it also has some other rather odd, and much studied, effects. Because the right eye is wired to the left side of the brain and vice versa, and because the brain's task of creating certain types of awareness also seems to be geographically constrained, some surprising things occur. For instance, in some subjects it was possible to place something, a written message say, in such a position that the subject could read it, without being aware they had read it.

That in itself sounds hugely peculiar, but observe what happens next. In one instance, a patient was shown a sign during an interview which read 'leave the room'. The person, unaware such a sign had been flashed at them, nevertheless got up and began to walk out.

'Where are you going?' the interviewer asked.

'I'm thirsty,' came the reply. 'I'm just off to get a drink.'

So, were they thirsty, or did they just think they were thirsty? Is there a difference? Clearly there is a difference between deep and described motivation here, and it's sort of unsettling, the ease with which consciousness can ascribe plausible, but ultimately irrelevant, causes to the body's actions. If one of the tricks the consciousness can pull is to make us think we're in control when we're not, well, where's that going to lead us?

In another case a woman who'd had the same operation describes going to get a green dress out of her wardrobe, only to find her right hand insisting on grabbing a red dress, and refusing to let go. As if another part of her brain had made a different decision, and as if information regarding this decision could not be spread about the rest of the brain in the normal way, the conflict could not be resolved. Indeed the woman needed help before she could get the hand to release the dress. Her own hand!

So, what was going on here? We begin to wonder if the consciousness isn't normally part of the process whereby competing urges are in some sense sorted out. Was the problem here that the woman couldn't decide which dress she wanted, or that she had decided, but wasn't aware of the decision? Was it that her consciousness hadn't been told, and so filled in the gaps, wrongly, and kicked up a fuss when the physical process outvoted it? I'm lapsing into loose metaphor here of course, but see the speculation that begins when we have a better understanding of the things a mind can do? It forces us to wonder what we mean by consciousness, when the illusion of an all knowing, all deciding central control unit is broken down. It forces us to play with our stories.

To move to an unfortunately common example: head injury can lead to massive changes in personality. People who before were phlegmatic, jovial, and compliant may overnight turn surly and uncooperative. It's not just a reaction to the stress of injury and recovery, but a noticeable long-term change in the type of person they are. So we might also say they are no longer their old selves at all. It is like, to use the common description, they have become a different person. (To a lesser extent the same thing can be observed of people under the influence of drugs.) The brain has changed, and so have they. This leads to problems for those of us who have been brought up in the tradition of thinking about a person as a continuous entity existing through time, of having an unchanging essence to which they are answerable. Which of the two personalities represents the true person? Is the question even sensible? Is the idea of a person itself something of a convenient, practical metaphor?

People with multiple personalities present an interesting twist. Some appear to be able to go through life with entirely separate narratives accumulating along the way. The inevitable gaps are effortlessly filled in with the most outrageous fantasies, and the subject swears they are real. Again, the consciousness seems to be remarkably good at filling in the gaps, ascribing stories and motivations quite independently of sensory inputs. Again, the idea of a single body possessing a single soul, a single essential identity, is knocked about a bit.

Blindsight is another provocative phenomenon. A patient claims not to be picking up any information from a particular part of the visual field, for instance if they are staring straight ahead at a computer screen, the blindsight portion will appear blank to them no matter what is on it. Yet their eyes are still working in the way of a normal seeing person, so the problem is not, as for the blindspot, with the incoming data. Interestingly, if you put a simple figure, say an X, on the blind segment of the screen and ask them to guess if it is an X or a Y, they get it right most of the time. In the same way, they will instinctively duck out of the way of an object thrown at them from the blind sector. (Beginning to sound fun, this psychology business, isn't it?) It's as if the patients can see, but are unaware they can see.

What does this encourage us to think about our consciousness? Perhaps that it is the result of specialised networks which don't so much deal with incoming data, as allow us to be aware of these dealings. It's a place where the reporting in occurs. So, although we might not yet wish to say that consciousness is just these connections and nothing more, we can certainly find good reason to believe that without these connections, there is no consciousness.

Consciousness, whatever it is, looks to be tied to particular physical states by a very rigid relationship. Metaphors of a free floating 'me', working independently of the physical workings of my brain, seem somehow inadequate.

Working our way through the list of disturbingly compelling afflictions, there is also the loss of the ability to recognise faces. Face recognition is something we take for granted, yet attempts to write face recognition computer software shows it to be a massively complex task; a task we happen to be remarkably good at. The same is true of forming sentences, interpreting speech, reading other people's moods, or maintaining our balance while walking. We must be careful not to overlook the wonder in the very familiar. Recognising one person from another should be as difficult as distinguishing between sheep in a flock, but so good is our programme for picking out small differences in faces, that we find we only need to see a person once to later have a fair idea they are familiar to us. (Some are far better than others at this game of course, a cause of regular embarrassment for the rest of us.)

When a certain part of the brain is damaged, however, the unfortunate sufferer will find the whole thing is suddenly a mystery to them. They can still recognise their house, their car, a piece of music, their partner's voice, shoes or trousers. But, until another clue is provided, they can't tell their lifelong partner from a complete stranger just by looking at her face.

Interestingly, when studied more closely, it appears that the people do in fact recognise familiar faces, in that when shown a photograph there are specific measurable skin reactions occurring. Yet like the blindsight patients, they don't know they know. Again, consciousness, the parcelling out of information at the level of awareness, which allows us to say we 'know' something, has broken down.

If we think a little further about what this breakdown means, we see that without conscious recognition, the information can't be used in any of the deliberate ways we associate with rational thought. Although the body may have 'recognised' the face of the wife, the sufferer, without conscious access to this information, can't now check the expression on the face to read her mood, and match it against his previous knowledge that she has just returned from the dentist. He can't choose a greeting appropriate to their relationship. Nor can he pre-empt her questioning with an apology for having forgotten to call in at the supermarket, slipping in a compelling excuse while he's at it. Consciousness, whatever it is, seems to be about making data available for this sort of deliberate manipulation, allowing us to indulge in sophisticated behaviour which instinct alone can not carry.

I could go on, but then there would be a danger of ghettoising this issue as the study only of damaged brains. Instead, let's turn to our own everyday experiences, to uncover more clues about the working of our ridiculously complicated thinking device.

We all know the situation where a person we are with has noticed something in the surrounding environment, but we, for all our squinting, just can't see what they are pointing at. 'Look, straight ahead of you. At the very top of the tree. See its tail. Look, a monkey.' So you stare at the top of the tree. Nothing, nothing, nothing … and then suddenly, as if by a miracle, the monkey appears, and you can't understand how it is you couldn't see it before. The data was coming in, but the picture itself was not forming. You hadn't managed to pick out the key details and build a picture of a monkey, or at least that picture hadn't been made available to your consciousness. (Your what? See how quickly we find ourselves thinking in terms of our little internal viewing creature?) Consciousness feels like a moment of realisation, of making sense of the world. A moment of connection almost. The disparate parts are suddenly put into a context that makes sense. It's like being back at school, having something explained to you that you really don't get at all, and then suddenly, for the lucky ones, there's the 'aaah' moment. We talk of things 'falling into place', and maybe this is a very good metaphor.

We also, in our everyday thinking, do an awful lot of filling in. We all have a blind spot, a point in our visual field where, thanks to the way evolution built our eyes, cobbling together a place for the optic nerve to exit, we receive no visual information. But unless we force an experiment upon ourselves, closing one eye and staring at a dot as we move it away from us, we remain blissfully unaware of this deficit. We don't see a gap in our visual field. Rather, in some hazy sense, the mind just makes things up, assumes that things that were there a moment ago are still there, that lines continue unbroken and so forth. In fact, many experiments show that perception is dominated by this sort of guessing. It would be truly inefficient to continuously sample and process data from a static environment, and so we just pick up the basics — edges, colours and the like — and sort of assume how the rest of it goes. Similarly, the flickering frames of a film, when sped up sufficiently, give the appearance of fluid motion. The missing bits are not apparent to us.

In part, this is why spotting the monkey in the tree is so difficult. We've seen enough tree stuff to assume it's all tree. In seeking out the monkey, we are forcing our brain to let go of those assumptions. Experiments show we are lousy at spotting where changes are made, unless we see the actual change take place. Items can be slipped in and out of pictures we are observing, and so long as the eye isn't resting on them at the moment of change, we don't know. As Daniel Dennett likes to point out, most of us are surprised to see just how narrow our range of colour vision is. (Hold a playing card off to the side of your head and stare straight ahead. How far around must you bring the card before you can determine whether it is red or black? It's surprising, isn't it?) We receive no colour information from our peripheral vision, yet experience the world as if it is continuously coloured. Walk into a room you are familiar with. Do you need to look around and take in the details afresh? Of course not. Instead the familiarity of a few visual clues allows you to access your memory and literally make up the rest. So it is that we trip over the chair someone has moved without us noticing ('But it isn't usually there!', you protest) or manage to offend by not noticing the vital new addition to the environment. ('How can you not have noticed the new curtains? They're a different colour!')

These observations should alert us to the fact that perception isn't as straightforward as we might assume. A lot of the detail we think we perceive may in fact just be the referencing of previous memories. This in turn alerts us to the fact that getting to the heart of memory might be a prerequisite to properly understanding consciousness.

On another tack, as a writer I'm particularly interested in the way words jostle their way to the front of my attention, whether I'm writing them down or speaking them out loud. I'm not the sort who likes to give prepared speeches. I'm far happier just talking off the top of my head, and when I do a curious thing happens. Part of my conscious self shuts down. I am in the moment as they say, caught up in the flow. Occasionally I may check in, thinking ahead to where the next point might take me, but beyond that it's a matter of kicking back and letting the words do all the work. So too when I write. I am not consciously choosing the words. Instead, the words seem to arrange themselves across my computer screen. (Although delightfully, as I write this, those very words have dried up. My self-consciousness has temporarily killed off the process.) Some writers or raconteurs like to use out-of-body metaphors for this process. 'I am not the source of my inspiration,' they might say, 'I am just the device through which it is channelled.' Here they seem to be distinguishing between themselves as their brain, and themselves as their consciousness. For my part, I prefer a less mystical representation. I am damned sure I choose the words, and am happy to take full credit for them. It's just I don't do it consciously.

The more you start to observe yourself in conversation, the more unsettling this notion of the unconscious communicator becomes. My consciousness is definitely doing something during a discussion (although the longer I am allowed to proceed uninterrupted, the easier I find it to zone out completely — boring even myself!) but I would make the claim that one thing it is not doing is choosing the words. This is not the same thing as saying that we can not consciously choose our words. Indeed, when the stakes are high enough, we may choose to do exactly this, self-consciously overriding the automatic mechanism. Look at the way a witness in a court case struggles to find the right words under cross examination, taking great care to say exactly what they mean. (Or so it is on television anyway.) As a writer, I have long and involved arguments with my own subconscious process, a ritual known as editing. On rereading my work, I carefully check it for flow and meaning, deliberately removing the excesses of my subconscious self.

Notice again how naturally we slip into talk of 'I', as in 'I override my subconscious', as if me and my subconscious are almost separate entities. It is tremendously hard to find a different metaphor, one that works to make sense of our world. That aside, the notion of editing, either after the event, or in a moment of high pressure self-consciousness, provides another way of looking at consciousness: as a protective self-checking mechanism. It's a little like a teacher on playground duty, watching the children play, only intervening when someone looks like they are about to hurt themselves, or when the teacher gets bored. We don't have free will, one wag remarked, we have free won't.

If writing and talking are interesting cases in point, how about reading? What do we make of the situation where we somehow let our mind slip off the task midpage, yet the reading still continues? You reach the end of the page, ready to flip it, your eye has scanned all those letters, your mind has made words and sentences out of them, a tremendously sophisticated task, and yet somehow, none of what you read has gone in. The crucial bit, the information, is missing. You have to go back and reread it all. This highlights the link between consciousness and memory again. Those things which we can't recall, which we have no working access to, if you like, don't feel like they are conscious acts. It's like snapping to, after a day-dreaming diversion, and realising you have somehow managed to drive another ten kilometres without remembering any of it. It's as if we should not think of consciousness as the thing that drives the action, so much as the thing that records it, and so makes the attendant data available for future use. But that's not how it feels, is it? We think of consciousness as a far more decisive, proactive force. The more we think about it, the more we feel as if we are being pulled in different directions. No wonder Descartes found it all so troubling.

This is all just throwing snowballs on the exposed tip of the iceberg. We could go on to consider what dreaming means, what it's like to be drunk, how we interpret the behaviour we observe in animals, how depression affects our sense of self, why fears are more likely to visit in the middle of the night, or how a child's consciousness differs from that of an adult's. With each new contemplation, this thing we variously call our consciousness, our mind, or our sense of self, becomes a little more complicated, and our intuitive models feel a little more inadequate.

This combination of scientific advance and steely introspection forces us to rethink our metaphors. We build models of the things we observe in the world, models which allow us to both explain and predict relationships. In the case of neuroscience, these predictions range from the effectiveness of pharmaceuticals to theories on education, behaviour control, child-rearing and rehabilitation following injuries. These models in turn must be interpreted by us; they must be slotted into our stories. In the case of the brain, the mind and consciousness, these are precious stories indeed. Stories about what it means to be a human being. As always, we need to be very careful to distinguish between the science, and the story part. Arguing about scientific concepts is helpful, a powerful way of motivating us to design new experiments, conduct new tests, and refine our explanations. Arguing about stories is more frustrating, lacking as it does a clear mechanism for resolving differences. It can be diverting and enjoyable, but it can turn dark too, fuelled by the deep conviction of our passions.

This takes us nicely from the world of cognitive psychology and neuroscience to the world of philosophy, and specifically the philosophy of the mind. Philosophers, by vocation, enjoy a good scrap: we shouldn't be surprised to find them embroiled in deep and twisted arguments. But remember Popper's insistence that they take a common sense approach, and concentrate on problems with real life applications, rather than those that are doomed to spin off into a tangle of meaningless verbiage? So what are philosophers saying about the mind? Are they building coherent cases, with real life, testable implications, or are they simply telling stories?

The basic question exercising the minds of the mind folk is this: when science has finished with its brain work (explained how networks are formed, attention is directed, processes are managed) will there be anything left still needing explanation? Of course we understand science will never be finished in its work, but this is a hypothetical argument. In some philosophical circles, the proposition advanced is that at its core, consciousness is unknowable. No matter where science goes in the future, there is an aspect of human experience it will not be able to explain.

This is a difficult argument to make of course, for science is tremendously hard to predict. Some problems which appear on the verge of being solved now may prove stubbornly intractable, and others which appear beyond us might yield to a single moment of insight. Maybe in a hundred years neuroscience will have a radically different model of the brain, and maybe it will still be working on refining our current hunches. Who knows? There is nothing in the history of science to make us confident either way.

But what if, the philosophers are asking, we had the answers to all the mechanistic, procedural questions surrounding thinking; would that be enough? Could we then stop and say, in the spirit of our blue sky explanation, that we now know what thinking is, or would that be to ignore the elephant in the corner of the room, which out of politeness we have all refrained from mentioning?

This debate has an interesting equivalent in one other area of science: in cosmology. Because the terms are more easily defined there, it might serve as a good analogy. On the one hand, there are those who believe that as we learn more and more about the nature of the universe, about what we mean when we talk of time and matter, about the birth of the universe, its dimensions, and ultimately its fate, we will move closer towards understanding 'what it all means'. On the other side are those who say that such a hope is futile; that the ultimate mystery, why the universe exists at all, can not be solved by the method of uncovering the laws of the universe, for there will never be a law which by its own definition *has* to be. There will always be the question, why this law and not another one? Indeed, why something and not nothing? (I am reminded of my childish collision with these sorts of mysteries, when the single word God was offered as an answer to them all. 'What was there before God?' I would ask. 'God created time,' came the answer. 'And before time, there was no "before", was there?' That sort of answer made my brain hurt. It still does.)

Nevertheless, we can see the two opposing instincts at work. The instinct to believe in the power of the human intellect to uncover truth, and the instinct to believe in the power of mystery to shroud the universe. Both are articles of faith: cultural and emotional tendencies. As yet, science has produced no plausible solution to the mystery of existence, and as yet the romantics have produced no plausible argument showing why a solution must be impossible. This is at heart a question of the capacity of the human mind, and it's not going to be solved in the abstract. While many scientists spurn the faithful and the mysterious, when we reach the outer edges of knowledge, the scientific quest itself becomes a mystical, faith-based endeavour.

What is the equivalent of the 'why something and not nothing?' question in the world of consciousness? Perhaps it is best introduced by the following reaction, which some of you may have already had, to all this talk of neuroscience:

'This brain science is all very well; it can help us to explain what the brain does when thinking occurs and it might even help us to equate brain states with types of thought, but that in itself can't explain what thinking is. You might one day be able to explain how the various patterns of brain cells are activated to represent in my head the colour yellow, but that doesn't explain why yellow looks the way it does to me. Why I have the particular visual sensation I have when I see a yellow object. There is something about thinking, the appreciation of qualities conveyed by the senses, that is intangible. When I look about me, I do not simply experience a complex train of chemical reactions triggered by the light quality of the visual clues. I actually see things. That seeing might coincide exactly with a certain network of neurons being activated within my brain, but that doesn't explain why I perceive it the way I do. Why doesn't seeing feel like noise to me? Why doesn't it have a taste? Why is listening to music not the same as having a rasping feeling at the back of my throat? What gives each of these recognisable sensations their particular quality? What is it about the way neurons interact that demands this particular type of feeling? Seeing, like contemplating, imagining, anticipating, comparing, feeling and abstracting, or being human, in short, can not be described in purely physical terms.'

Does that resonate with you? If it does, then you are in good company. There seems to be something fairly solid standing in the way of a jump from mechanical processes to the quality of our sensations.

Certainly my every instinct rails against it. But are these instincts any good? What exactly is the nature of these mysterious experiences, which we feel so strongly, and talk about so freely, but which we find so hard to define? Some philosophers, Daniel Dennett being one of the more extreme and therefore well-known, have argued that these experiences are in themselves nothing more than a metaphor, a way of talking. Once we have successfully modelled the physical elements of the thinking process, there will be nothing left to explain. Dennett claims that thinking is at heart nothing more than a mechanistic process; albeit the sort of subtle, complex and glorious procedure that only a machine of 10^{15} (a million billion) connections could manage. His claim is that the mysterious way we think about thinking is a result of the mysterious way we speak about it. We have talked ourselves into a corner, using words to convince ourselves of things which, upon closer examination, we do not actually experience.

Dennett's model of consciousness, although by his own admission undeveloped, points to the way he thinks a change in perspective might one day strip this debate of its central point of conflict. He has called this model 'fame in the brain' (amongst other things) and he claims in time such models will become sophisticated enough to convince us our thinking about consciousness can be reduced to thinking about brains; just as we have been able to reduce our thinking about life to thinking about chemistry.

This is a compelling metaphor. Once we believed life was an essential force, breathed into the animate by some greater spirit, a thing well beyond our capacity for knowing. Now we think of life in terms of DNA, proteins and cells. He predicts that consciousness will some day go the way of life; there will be an 'aaah, now I see, it isn't such a mystery after all, silly me' sort of a moment. Lined up against him are those who claim that this isn't a life-type problem, but rather a universe-type problem. For them, 'why do sensations exist?' is as mysterious as 'why does the universe exist?' No matter how well mechanical models are built, they will always falter at the final stage. For they will not be able to explain how a pattern of material excitation itself can translate into the things we experience, the things we call thinking, feeling and knowing.

Like many philosophers of this school, Dennett has a keen interest in artificial intelligence, and we might see that as the practical problem which gives philosophy its licence to explore. For the artificial intelligence programme presents us with some arresting possibilities. There was a time not so long ago when the very notion of thinking machines was preposterous. But today, with computers effortlessly whipping us at chess, adjusting the handling characteristics of our cars on the move, translating passages of text from one language to the other, and linking our most esoteric questions to relevant answers on Google, the argument is not so much whether machines think, as whether their thinking is anything like ours. And this is the sort of question that tends to galvanise the imagination (more than it should, I will later suggest, but one step at a time). If computers continue to progress at the current rate, then won't we, at some stage in the future, have machines that are capable of befriending us? How do we feel about that? What about a machine that is so good at 'thinking' that we can't tell it from a normal human being? It says the things we expect a person to say. It has a personality, exhibits shyness, fear, excitement and confusion. It tells us it feels real, it feels conscious, and it finds this consciousness deeply puzzling. It fears its own demise, and can match you existential doubt for existential doubt. How would we treat one of those? (This appears to be a deeply cultural question, one which is answered differently in Tokyo than it is in London for example, and that should warn us that we're stepping into storytelling territory.)

The possibility of a 'thinking machine' provides the basis of what philosophers refer to as the zombie problem. Once a machine achieves the level of sophistication required to produce human behaviour, do we have to assume it has also achieved a level of consciousness? Or, is there the zombie escape? Can we argue that although the machine does remarkably human-like things, it does not experience human-like thinking? In other words, can we reasonably say it doesn't really think, it just gives the appearance of thinking? Our instincts certainly lead us to the second explanation. We are more likely to feel a pang of guilt pulling a fish from the ocean than shutting down our computer. And we do not base this on processing power alone; rather we are able to imagine that the fish, being alive, has feelings, like ours, and is able to experience pain, like we do. And the computer is a machine, which makes it different. That's just how it is, we assume, and to us the difference matters.

But hold on, say the anti-zombiests. Have we thought about this carefully enough? What makes us so confident that a machine could really do human-like things without also experiencing human-type feelings? These objectors are following up on a very provocative and powerful hunch, the hunch that what we interpret as sophisticated 'sensations' are actually just a collection of far simpler reactions, which the human brain, with its facility for language, and therefore metaphor, weaves into a rich tapestry of associations. Perhaps this associating, or networking, of brain data, is what physical and emotional sensations actually are: a buzzing, reverberating feedback of memories and experiences.

I'm using some pretty vague language here, which should alert you to the fact that as yet this is less a research programme, and more the beginning of an idea: the sort of punt that might direct future experiments in psychology, and future algorithms in artificial intelligence programming.

I just tried a little experiment of my own which might help clarify the concept. On my desk, below my computer screen, is a vivid yellow highlighter. While typing I am aware of its presence, for the colour will hardly be ignored, but I am not aware of it in any detail. Without focusing on it, I find I know it is yellow, but when I turn my attention fully to it, the yellow I see is somehow different. Sharper, more clarified. As if my previous notion of the colour was nothing more than a vague accessing of memories of yellow, a noncommittal canvassing of a range of possibilities. When I look, the range is narrowed down, and previous memories and associations of exactly that sort of yellow resonate more strongly.

This raises the possibility that the sensation we get when we view a yellow object, the feeling of seeing yellow, this supposedly indefinable quality, has something to do with accessing memories of other yellow things. This leads to the very important possibility that the quality we call 'yellowness', the sensation of seeing yellow, is not an irreducible thing, but can be further broken down into component parts. One might argue that if these component parts are memories of other sensations, then no reduction has taken place, but could it be that the first time we experience a colour, with no memory to refer to, the sensation does not have its defining characteristics? We don't remember what it was like to see things for the first time as a child. What if, at first, it wasn't 'like' anything, in the philosophical, elusive subjective quality sense? What if sensations are in themselves neural networks, and the development of consciousness, tied as it is to the use of language, metaphors about how groups of thoughts and memories can be arranged, is the process by which sensations are also developed?

This leads to questions about what memories are made of, and what the sensation of memory accessing really feels like: whether there is a generic accessing sensation which in turn can be used to build the distinctive quality of sights, sounds, smells and contemplations. This is pure hypothesis-on-a-whim, something that just popped into my head this moment. I am not saying that is what sensations are, I am just trying to show that assuming sensations must be in some sense irreducible and resistant to analysis might be a tad premature. There may be ways of finding this stuff out: so why are some arguing that in principle it simply can't be done?

Try imagining in detail the face of someone you know very well. How sharp is the picture? Sharp enough that you can place the mole you are sure they have? Sharp enough that you can tell me the colour of their eyes? The pattern of that colour, around the pupil? How well can you describe the style of their hair? Its length? Its exact colour? When they smile, which teeth are visible? Could you pick their nose from a line up of photographs (unfortunate phrase that)? Eventually we find our picture loses the necessary level of detail (this happens a lot more quickly for some than others: I'm useless at it). Yet we persist with the illusion that when we imagine something, we literally see a picture of it. It just feels like we're seeing a picture, the same way it feels like we're seeing yellow. So again we find the possibility that a supposedly irreducible and objective sensation — seeing a picture inside our head — can be broken down into more basic sensations, the feeling of accessing certain networks of memories. This might lead future researchers to discover an answer to that most vexed of questions, why does the sensation of receiving and piecing together information from the world feel the way it does?

The very possibility of such research leads to the anti-zombie hunch: that the sort of information flow producing human behaviour must also by definition produce human feeling (here I mean sensation, rather than emotion), because that's all feeling is: the flow of data. But it's a hunch roundly dismissed by some leading philosophers, and most lay people when they are first exposed to it. Why? Because we can't escape the feeling that whatever the true quality of the experiences we have, and however misguided our conceptions of it may be, it's still 'we' who are feeling, and in some way this must be explained. No matter how good those explanations, we like to think that we as individuals will lie at the heart of the mystery, not some ultimately explicable neuronal network.

In a way, that is as far as the zombie debate takes us. Some claim they can imagine fully-fledged zombies, machines doing all the thinking things we humans do: processing incoming data, accessing memories, relating them to programmes of intention, forming plans of action, testing hypotheses in the abstract, acting upon a chosen path — yet without having any of the associated sensations. Others claim not to be able to imagine such a thing at all. They claim that such things are impossible, for the processing of such information, and the feeling we get from doing it, are one and the same thing. Can we ever imagine this sort of debate being resolved? It appears problematic; for the trick of the scientist, designing falsifying experiments and using the outcome as a tried and true means of adjudication, isn't available to us for now. How could you test whether the zombie really had conscious feelings? We could ask, but part of the deal is that it has to say it does, just like we do.

Right there is the other question the zombie problem raises. How do I know you are conscious? How do I know everybody in the world but me isn't just a zombie? I don't, of course. You might all be cleverly programmed liars. But I have excellent, pragmatic reasons for believing otherwise. I wish to avoid the crushing sort of loneliness that would come from thinking any other way. I have an instinct for meaningful connection. I choose to believe that others share my experiences and feelings, and that I can therefore be part of a community of sentient, caring and emotional creatures. Whether I ever chose to admit machines into this community probably has more to do with this sort of instinctive and self-serving storytelling than it does any sound and testable scientific basis. For now, at least. The anti-zombiests urge us to follow through with the science, because it may one day explain, in purely physical terms, what consciousness is all about. I don't know if their optimism in this is misguided — time will tell — but I do find something odd about those who argue that we shouldn't even try, because material explanations simply can't explain consciousness on principle. What principle would this be? We can perhaps get nearer the heart of this darkness by considering another thought experiment that has been proposed in support of the anti-materialists.

The experiment is known as The Chinese Room, and purports to show that material reductionism can never work, because there is a clear distinction between perception and behaviour. In other words, there is a quality of consciousness which exists independent of the physical workings of the brain. John Searle is the philosopher responsible for this one, and he and Dennett have argued long and hard about it, each claiming that the other is missing the vital point. It works something like this:

Imagine a weird and wonderful room. There is a small slot in one wall, through which messages can be passed, and inside there is an elaborate machine consisting of pulleys and levers, and a thick, unambiguous instruction manual. Also in the room is a human operator. Outside the room is a speaker of some language, we'll say Cantonese, and she is told that inside the room there is another fluent Cantonese speaker, who wishes to engage in a conversation by way of written notes. So, our Cantonese speaker outside the room writes out an initial message and passes it through the slot. The person on the inside retrieves it, but in fact they don't speak a word of the language. It's all just mysterious scribbling to them. Never fear: they go to the instruction book and match the symbols one at a time to the instructions, which tell them which levers to pull, and which buttons to push. After suitable whirring and other magical noises, the machine spits out its instructions, telling the poor illiterate what to put on their paper. They do this, and pass the message back out. The reader receives it, and indeed it is a suitable response in perfect Cantonese.

So it goes on. The Cantonese speaker feels they have made a new friend. They are fully engaged with this witty, stimulating, conscious being. They can hardly wait to meet them. Yet inside there is nothing to meet at all, just some poor schmuck pulling levers, who has no idea what they've been talking about. See? says Searle, it is possible (in this imaginary world at least, so possible in principle) for us to create the appearance of conscious thought without any conscious thought being present. The two things are not the same at all. So we can have zombies, for this is one, surely. The fact that we can have zombies in turn demonstrates that there is a difference of some sort between the functional processes associated with consciousness — the brain states if you like — and consciousness itself.

Is this in fact the clinching argument Searle hoped for? To answer that, we have to look at the proposed situation a little more carefully. Consider what happens when we fill in some details.

The Cantonese speaker is having a bad day, and as an opening gambit writes, 'Come out and show your face, or I'll burn your building down.'

Okay, now it seems to me we have an interesting problem. What should the building respond? Here are some options:

'Why do you want to burn me down?'

'Having a bad day, huh? Want to talk about it?'

'You set light to me, and I promise you I will kick your arse good and proper.'

'I'm very shy. I don't want you to burn the building down, but I'm very nervous about coming out.'

'Sure, go ahead, burn it to the ground. See if I care.'

For The Chinese Room to work, we need to know, at least in broad brush terms, how it will go about choosing one of these answers over the others. I don't mean how to choose the wording, but two other far more important things. How do we work out what the intentions of the Cantonese speaker are, and what in turn are our own intentions? Do we take the speaker seriously, or are they just joking? Perhaps it's just bravado. We need to work this out early on. How do we feel about this? Are we spoiling for a fight? Are we looking for a friend? Are we just bored, and seeking stimulation? Perhaps we're feeling suicidal in which case the threat takes on a rather different hue. Meaningful conversation of this sort appears to require something rather special: both motivation and interpretation of motivation.

We may argue that the room is programmed with some intention in advance — say making a friend, and that it then uses a series of preprogrammed strategies to pursue this goal. E.g. if I encounter aggression, respond first with an offer of friendship and understanding. If this leads to more aggression, back away and minimise my responses, and so forth. This sounds like it will get impossibly complex fairly quickly, but it's a thought experiment so we're allowed to assume a brilliant programmer. It's not impossible to imagine short cuts for categorising that might be built in, and some tricks for steering the conversation back to familiar territory. After all, we do the same ourselves when we engage in conversation. We categorise what we hear, and try to work out what the other person's perspectives and objectives are. But this brings us to an interesting point. The machine is going to have to be a very special sort of programme, one that is not too dissimilar to the programme we ourselves pursue, when engaging in conversation.

This leads to a possible answer to The Chinese Room dilemma. It may be true that the person inside the room, as an insignificant part of the machine, is not conscious of the conversation, but it does not immediately follow that the machine itself doesn't know what's going on. If consciousness is, as the materialists suggest, nothing more than the manifestation of a certain complexity of information flow, then when the machine has reached this point, it too becomes conscious. We can certainly see how massively complex the machine would have to be. So why don't we therefore confer consciousness upon it? This argument may appear ridiculous, but we still have to accept that it has taken things back to our previous zombie conundrum. Until we have a tighter definition of what it means to be conscious, of what it means to understand, to want, to feel confusion or fear, we can't reasonably know whether the room itself has achieved the level of consciousness we are trying to dismiss from the equation. (We could ask it, but the room will reply, in perfect Cantonese, that it does, for these are the rules. The person outside the room must be fooled.) So we can't rule out 'but the machine is conscious' as the answer to The Chinese Room problem.

Searle doesn't think this is the point though, and has suggested reducing the problem to one of a person standing alone in a field, who by dint of a massively effective memory, has learnt by rote appropriate responses to a range of possible Cantonese statements. The stranger approaches, says some stuff, Ms Photographic Memory answers, it goes on like this for a while, and then the stranger walks away, having made a fulfilling, conscious connection. The woman in the field, though, has no idea what really went on. I've experienced something like this when travelling round France. My French is somewhere between poor and non-existent, but with lots of smiling, nodding, and 'Oui, oui,' plus the occasional 'C'est bon' and always a 'Merci' to close, I was apparently still reasonably fun for a certain type of indolent to chat to (or perhaps they were trying to pick my pockets).

Even without the language complication, there's no doubt we do sometimes zone out of conversations, producing words on autopilot, investing little or no meaning in the proceedings. Have a look at the dialogue in some of the great absurdist plays, the opening of *The Birthday Party* by Harold Pinter perhaps, to see how interpretation, intention and utterance can be uncoupled. So, if we reduce the room to this example, Searle has a point. It is, in principle at least, possible to hold a conversation without investing it with what we would think of as conscious thought.

However that's just for the simplest form of conversation. If the French speaker, or the person you are zoned out from at the dinner table, asks a question requiring a complex response, don't you then by definition, have to engage your consciousness? And isn't the criterion for The Chinese Room that the native speaker is made to believe there is a conscious person on the other side? Doesn't that imply a certain level of complexity?

Again we find ourselves in danger of entering a circular argument. The first problem with both the zombie and The Chinese Room is the lack of useful definitions. Consider a point not often made but which appears vital. I don't think many people would pass the zombie test, or The Chinese Room test for that matter. Put a fluent Cantonese speaker inside the room, but tell the person outside it might just be a cleverly programmed machine they're talking to, and how long before the outsider begins to have their doubts? How many of us are actually capable of convincing a probing paranoid mind our answers are genuinely thoughtful? If the criterion is 'can fool another person' then it's a poor test, for that's too loose a definition to help us in our probing. What sort of person do we have to believe it might be? What say we were told that inside is a fully conscious three-year-old? I think a fairly simple and fully unconscious machine could pull that off. How about if we're told it's a drunk, a fool, or someone who just takes great joy in being irritating? The point is that we don't confer consciousness on others because of the way they behave (although we do make judgements about attentiveness). We don't test them out. We just believe they have what we have, because we know they're built through a similar process, and as I've already said, it's a tremendously useful belief to have. There is no test for the zombie or the room to pass, for consciousness of the type explored here is not binary; you don't either have it or not, rather it comes in many shades, now fading, now sharpening.

The other problem is trying to pin down at which level of complexity consciousness kicks in. Introspection tells us that for humans at least, such a thing exists. Some conversations I can leave pretty much to my subconscious, but as things get complicated, I need to be more proactive in making interpretations, relating them to context, reading accompanying body language and the like. Where that level is depends upon my skill as a conversationalist. There is no definable cut-off point, and even if there was, that's not the rub. Searle is saying, for at least some level of communication, a given set of responses can come wrapped in meaning and intention or not, and it makes no observable difference to the style of communication occurring.

On this matter, I think he is indisputably right. I also doubt it makes the point he hopes it makes, for I don't think it tells us anything useful about consciousness. It certainly doesn't establish that there is more to thought processes than the accompanying physical actions.

Think about some of the possible models of consciousness. What say we go with the idea that consciousness is awareness, the process by which experiential data is made available to the memory for further use and manipulation? In this case, at some level of complexity, in order for the conversation to hold together, we are forced to become aware of the things we are saying, in the sense that we need to make them available in our working memory, so we can put them in context, relate them to our superobjectives and so forth. If, according to this model, the feeling of being conscious is nothing more than the excitation of these awareness networks, then zoning in and out of conversation is analogous to bringing these networks in to or out of play. If the room is to use a similar technique, it too will experience consciousness. If it isn't forced to, if its set up is so fiendishly clever that it can achieve on autopilot what you and I must concentrate to manage, then so what? The 'something extra' that consciousness is supposed to be is not indefinable in this case, nor can we be sure it is something other than the physical process required to allow humans to achieve the trick of attention. This model of consciousness may not be right. It may be nowhere near the truth. Yet while it, and other models like it, remain possibilities, The Chinese Room hasn't established the non-physical nature of consciousness at all.

Okay, if your head is hurting here, you're not alone. Philosophers who have devoted their whole lives to these sorts of questions are still arguing about them. Let's take a deep breath and see if we can go back through the implications of The Chinese Room slowly and quietly, to make sure we're not missing anything important.

Searle thought, because the person inside the room did not have to understand the conversation, that this showed that consciousness and appearance of consciousness, must be different things. Although we can argue that the room itself may understand what is going on, this is a distraction from the key point. For Searle wants to make a further claim: that therefore consciousness must be something more than the physical processes (the neural networking) associated with it. Unfortunately for Searle, this point doesn't follow at all. I readily concede that I can give the conversational appearance of attentiveness without attending, but when I do attend, when I do put my mind on the job, I suspect the actual physical chemistry of my brain changes too. If this is the case, it could mean that this change, the particular sensation of making data available for real time, motivated manipulation, is exactly what consciousness is. It may not be. The jury is still out. But it is in principle quite possible, and in this case The Chinese Room establishes nothing.

The last problem I want to consider when we talk about consciousness is that we are in fact talking about two separate things. This causes a great deal of confusion. Ask someone why they object to consciousness being reduced to the mere physical workings of the brain, why they are still with Descartes on this one, and they might say something like, 'Because when I watch a stunning sunset, or when I describe it later to a friend, I am the one doing it. And you still haven't explained what this I is. Why do these things feel the way they do, to me?'

That's fair, it hasn't been explained yet, although the leap from 'hasn't been explained' to 'can't be explained' is the same one Michael Behe makes in his irreducible complexity argument, and as we've seen, in that form it's logically flawed. Let me draw your attention to another aspect of the above statement. There is an important category difference between watching a sunset, taking in the glorious colour, and thinking about the sunset, choosing the words with which to describe it. They're two different styles of consciousness. There is taking in information in a detached, unaware, subconscious way, and there is attending to this information, making it available for contemplation and communication. The Chinese Room dealt with the second type. It tried to show there is something special going on when we become aware of the conversation, understand its meaning, and its relevance to us. As we've seen, that may be true, but the thought experiment by itself can't establish this. What about the other side of consciousness, the fact that there are sensations associated with things? That cold just feels cold to us, that red things look red. What does this type of consciousness imply? Perhaps the mystery is here, at the level of raw experience, rather than at the level of sophisticated information manipulation. The reason we're not explicit in this I think is that we're trying to ring-fence human consciousness, yet dogs respond to smells, sounds and sights in ways that would lead us to imagine that they too 'experience' things on some level. It is harder (although by no means impossible, look at Descartes) to convince ourselves that while we actually see shapes, the family pet only responds to them, in an automatic, non-conscious way.

One last thought experiment tries to address this idea of sensations directly, and is known as Mary the Colour Scientist. Again, many people find this a convincing argument, but as with The Chinese Room, I'm guessing they're just seeing what they want to see. This one comes to us courtesy of the Australian philosopher, Frank Jackson. It goes a little like this. Mary is a brilliant scientist who has been able to interact with the world in a full and normal way, with the exception that she has never been able to experience colour. (Who cares how, it's a thought experiment. Perhaps she is made to wear black and white conversion glasses her whole life.) So she can get all the theory, stuff about light waves, about associations, about how light travels, how the cones of the eyes sense it, how the signals are sent into the brain, the whole deal. But she never actually gets to see it herself. Then, one day, the glasses come off and she sees the blue sky for the first time. Despite all her knowledge, the actual sensation of seeing blue comes as a shock to her: it is crucially different from an abstract or her intellectual knowledge of blueness. It has an intangible quality. We can't reduce conscious experience to knowledge of that experience. Ergo, consciousness is in this sense intangible.

Again, the argument is initially appealing. Sadly, for those who hope this is the clinching case in favour of mysticism, its construction is flawed. Just very briefly (because chances are you're tiring of all this abstracting) again our definitions are a little loose. What is Mary allowed to know, and what is she allowed to do to get the information? She can't take the glasses off, of course. But is she allowed to find out how the neural networks within our brain represent colour (if indeed this is a fair description of what happens)? If so, is she allowed to manipulate her own networks to represent this experience? Presumably not, for this is cheating. It's the process of seeing colour without the initial stimulus. If this sort of cheating is not allowed, what's the deal with us saying Mary is allowed to know everything about colour? We're actually saying that Mary is allowed to know everything about seeing colour except what seeing colour is like. If they're the rules then yes, the colour blue will come as a surprise to her. It will include new information, but it's new physical information she has been banned from collecting in the past. It's hardly evidence that colour perception is more than a physical process. It may be an argument that knowing about colour requires experiencing it, but if by experiencing it we mean nothing more than the stimulation of certain established networks in our brain, then we're hardly demonstrating knowledge has a non-material component, or is in any important way different from experience.

Another fascinating possibility which I quite like is that Mary, on being shown blue for the first time, doesn't experience blue at all. The assumption that blue will appear to her as it does to us is a brave one. For think back to the example of the yellow highlighter. What if colour appreciation develops gradually over time? What if seeing blue is indeed a sensation of association, one that builds up over our lives, that is in the end the feeling of accessing the memory of other blue things? Some work with the surgical reversing of colour-blindness gives potential support to this hypothesis. Again, with thought experiments, you can only go to the places your assumptions lead you, and as with zombies and The Chinese Room, Mary proves only that some people are deeply committed to keeping our consciousness fenced off in a 'too mystical to contemplate' zone.

Okay, enough of this geeky philosophy stuff (actually I quite enjoy it, and have been assuming you do too). Where has it got us? Well, I hope to some place where we can re-examine the roles of science and story again, because otherwise this diversion is entirely unjustifiable. I have argued all along that science can't by its nature dictate the story to be told from it. Think again of our child in the sandpit with the Lego blocks. In this case, as with evolution, we are faced with a curious reversal. For here the story tellers (anti-materialist philosophers) are telling us where the limits of science must by definition lie. We can never know what consciousness as we experience it really is, they argue. For this thing is unknowable, and beyond modelling.

The zombie argument, The Chinese Room and Mary the Colour Scientist are all attempts to convince us that scientists will never, on arguments of principle alone, be able to explain the mind. It's just beyond us, apparently.

All these arguments fail for similar reasons. They are circular. It appears we can only establish something is impossible by slipping a sneaky assumption into the set-up, and hoping it isn't noticed (we assume for instance that the room itself can't be conscious).

The problem here is our old friend the Platonic truth. If there are indeed knowable truths, we haven't demonstrated a method of reasoning our way towards them yet. We always have to fall back onto articles of faith: that the world is bound by consistent and simple physical laws, for example, or there is a sacred text which contains all truths, or the consciousness is a deep and unknowable mystery of the universe. Nobody has demonstrated a method by which any such truths can be established, which is why it would be a jolly good idea to get out of the habit of using the word truth altogether and admit that what we're really talking about is usefulness.

For what we have stumbled upon is a method with which we can build models that, in simple circumstances, make accurate and verifiable predictions. This method, although fuzzy in its definition, is what we call science, and it's tremendously useful. Because we can judge predictability objectively (some models make measurably broader predictions, or exhibit more accuracy in them) science is able to progress. This is not to say that there are no boundaries. It is clear that much of the world we experience is far too complex to be modelled by our current techniques. Ultimately, the limitation of our technique may well be the limitation of the human brain: biology may hold us back in our understanding. But there doesn't appear to be any good way of predicting where these limits are likely to lie in advance. We just have to keep modelling, and see where it takes us. Maybe one day we'll find a self-generating law which explains the existence of the universe, or maybe no such law exists. Perhaps it does, but it will forever be beyond the capabilities of our brain. Can we ever know which of these three is correct? Well, if we find the self-generating law, we'll be there. Until then, to have a belief either way is to make a leap of faith, not science.

This faith leap will in turn be partly driven by its usefulness to us.

For the scientist, the belief that the universe will continue to yield up its secrets is what motivates further enquiry. In other circumstances, pulling down a veil of mystery allows us to build the stories that make life such a satisfying experience. Some scientists like to argue that the acceptance of mystery is always and everywhere a bad thing, a celebration of ignorance over reason. Against this I would argue that there is something awfully killjoyish about insisting that the gaps in our present knowledge should not be garlanded by the beauty of story. Later in this chapter I'm going to show why even the most obdurate rationalist still falls back on stories to get them through the night.

That some continue to claim that the mind is by its nature unknowable, that consciousness is not the end product of a physical process within the brain (what Francis Crick admiringly termed 'The Astonishing Hypothesis') alerts us to the fact that for them even the possibility of such reduction threatens their most sacred stories, just as the evolutionary model once troubled sectors within the church. For them the very notion of conscious computers is not just beyond our current reach, it is also somehow repugnant. They (and it is a 'they' which I suspect represents the great majority of us) prefer to cling to the Cartesian divide. Thinking is a thing which thought itself can not explain. This is not an altogether crazy claim. Mathematicians tell us that the description of any system requires more information than the system itself contains.

Rather than heading off into the nosebleed world of abstract mathematics, though, I am much more interested in looking at the emotional quotient of this belief. What is it that frightens us so about the possibility of consciousness being explained? I would suggest it is to do with the way two of our most sacred myths are constructed. I speak of course of the two big Ds, Death and Decisions.

Mortality is a pesky problem, no doubt about it. No matter our creed or personal experience, the idea that one day our lives will just end takes more than a little getting used to. There is of course the brave philosophical claim that it is only through a contemplation of death that life makes sense to us, but that's always felt a little like bravado to me. For in that moment when we honestly contemplate our own non-existence, I suspect it's something of a stretch to find much that is comforting. The philosopher Thomas Nagel put it this way:

> There is something that can be called the expectation of nothingness, and though the mind tends to veer away from it, it is an unmistakable experience, always startling, often frightening and very different from the familiar recognition that your life will go on only for a limited time.

Or to use the slightly more down-to-earth epitaph:

> Here lies Jones, and he's none too pleased about it.

Because sitting around feeling paralysed by existential angst is not generally considered an excellent way of passing a life, we have to somehow deal with the fact that we die. One popular approach is to simply ignore it. Witness the growth of the 'forever-young industry' which will pinch you, pull you, dye you and deny you in the name of a youth that never fades. The aim, I suppose, is to keep our minds off the inevitable, and like any approach, it can only be judged by the way it makes you feel. At what age exactly do bicycle shorts become undignified?

Another approach to the death problem, and one with a far longer heritage, is the 'we don't really die' angle. We are in some sense immortal; there is a version of life after death. On the basis of its popularity, we must assume this view of the world has much to commend it. Just as the evolution model threatened the conception of a deliberate, interactive, hands-on creator, so too the 'consciousness as chemical and electronic happenings' model threatens our embrace of the afterlife. For if we, our thoughts, fears and aspirations, are nothing but the by-product of the brain machine, then when that machine stops working, there is no alternative 'we' to escape this unhappy state of affairs. Remember how important the soul was to Plato's notion of immortality, and how hard Descartes also pushed for the separate non-physical essence of self.

There will of course be a way around this conundrum. Just as flexible thinkers have had little trouble marrying evolution and religious belief, if you really want to believe in an afterlife, no physical model of consciousness is likely to demand you give it up. But the evolution debate is instructive in reminding us just how uncomfortable the transition is likely to be. And a physical model of consciousness is nowhere near as advanced as the evolutionary model of life. It remains for now just a hunch, an open possibility. Our fear of death is surely one of the reasons why some attempt to shut it down.

The other reason is even more vital to our sense of self. For the one belief we all carry with us, no matter how rational we claim to be, is a belief in our own free will. As with the death debate, this little conundrum has a long and distinguished history.

A child approaches her parent.

'God knows everything, right?'

'Yes he does.'

'So he knows what I'm going to do?'

'That's right.'

'Even before I do it?'

'Yes, he knows everything.'

'But you tell me I should choose to do the right thing.'

'You should. God wants you to.'

'But he already knows which one I'm going to choose. So I'm not really choosing, am I?'

'It's a mystery.'

'Oh, I see.'

Interestingly, the problem doesn't go away if we substitute an unbelieving, scientifically minded caregiver:

'Mum?'

'Yes dear.'

'You know how, at the super-quantum level at least, the world is physically deterministic?'

'That's correct dear. The outcomes you observe are the inevitable result of the circumstances directly preceding them.'

'And I'm just made up of physical stuff right, there's no weird soul thing?'

'Absolutely not. Wash your mouth out.'

'So how come, when I do bad things, you get angry at me, even though it's just physics happening?'

'It's not my fault that I get angry dear. That's just physics happening too.'

The choice, it seems, is between a contradiction and an absurdity. Or, as with death, we can just ignore the problem, which in this case seems somehow more sensible.

Free will, the belief that we make choices as we move through life, is an integral part of the human package, even if historically the commitment to it has varied. Some cultures make a lot more of fate than our own, which tends to push the 'choose right and be happy' button fairly hard. No matter what our upbringing, at the heart of every human interaction is a crucial assumption: that the person we are dealing with has some degree of control over their actions, and hence can be held accountable for them. Our legal system demands this belief. So does the satisfaction we derive from facing hurdles, making wise decisions, committing to projects, seeing things through. We reward loyalty and punish infidelities, we make plans, we appreciate sacrifices. To be human is to behave as if we are free, and to judge the behaviour of others by the same criteria.

So what if we aren't really free at all? What if our freedom is of the type enjoyed by a chess-playing software package, apparently able to choose between millions of available tactics, but in fact following the dictates of a cleverly constructed algorithm? Or of a single drop of water, its buffeted path too complex to model, yet somehow inevitable? What if our consciousness is the passive by-product of our path through life, rather than the driver of it, a marvellous device for attributing motivation to action after the event? Wouldn't that sort of knowledge just destroy us?

In the parlance of philosophers this is the difference between free will and apparent free will. Perhaps we feel as if we are free, but in fact we are not. We are just machines, crunching through a programmed existence. Nobody wants to know this, do they? Do we oppose physical models of consciousness because instinctively we sense we have too much to lose? Let that knowledge genie out of the bottle and mightn't its gift turn very sour indeed?

Actually I think this is a false fear, one we've been lured into by our Platonic thinking. For, if we loosen the strings a little on our definition of truth, then free will stops looking like such a big issue.

Once we stop thinking of absolute truth as something out there, demanding its own pursuit as part of the noble human quest, another possibility immediately arises. For what if it is far more sensible (and indeed consistent) to see truth as purely a function of utility? And by this measure, before we ask of a belief 'how true is it?' we need to ask 'What is the knowledge for?' Having a purpose allows us to judge the so-called 'truth' in terms of its value to us. This sounds like a sneaky sort of squirming out from beneath the question, I know, but before you dismiss it I challenge you to provide an alternative definition of truth which doesn't simply lapse into mysticism.

Let's wind right back to the opening chapter, for now we can see where all that set-up was leading. Ideas about the world are an invention of the human mind. What else could they be? Why do we, both consciously and by way of programmed instincts, bother with these ideas? — in order to smooth our path through life, to make it more fulfilling. Thus we have ideas about the physical relationships of the world, ideas which allow us to predict outcomes, avoid threats and construct new technological solutions. And we have ideas about our place in the world, about our life's meaning. We tell stories about our own motivations and the motivations of others. We form opinions about right and wrong.

What we must not fall for is the trap of believing that one type of idea, the predictive, scientific one, contains more truth about the world than the other. For this requires, amongst other things, a definition of truth which is not viciously circular. (True things are those which provide accurate predictions; therefore those models which most accurately predict the world are most true.) Scientific models have been constructed for a very good purpose: to provide predictive information about the world. This is an excellent idea, one of the best we've ever had, and we should celebrate it. But there are many times where scientific knowledge simply isn't useful, because it's not prediction we're primarily interested in. Free will is a superb example.

Ask yourself this. What is the difference between actual and apparent free will? Consider a person who says, 'It feels like we have free will, but actually it's all just an illusion. In fact we're just machines doing what the laws of physics demand of us. Our lives are going to work out the way they're going to work out, and there's not a goddamned thing we can do about it.'

Apart from a frisson of sci-fi excitement, what exactly does this statement yield? As far as I can see, they're making only one substantive claim: that if we were to replay life's tape, you and I, faced with the same circumstances, would make exactly the same decision, every time. We would have no choice. I find that a tremendously boring difference, for the simple reason that we don't ever get to replay the tape. Time, for reasons physicists still worry about, only travels in one direction, and under those circumstances feeling as if we have free will and actually having it are exactly the same thing.

Even if the difference were more substantial, I'm still pretty sure I would ignore it. This is a good test for your faith in the scientific model. Imagine a future where science has demonstrated without a doubt, in the strictly predictive sense of its methodology, that we humans are simply machines, slowly winding down our inevitable interaction with the physical world. Imagine the automaton model proves to be a better predictor of future outcomes. What would you do? Fall down in a heap of despair, say a teary goodbye to your old, driven self and lose the will to live? Or say, 'Actually that's a very silly thing to believe, for it will encourage me to fall down in a heap of despair, and make my interactions with others quite untenable. I rather enjoy my non-fallen, interacting existence. So I will dismiss your finding as most unhelpful, and continue to live my life as if I have some choice in the matter.' For even if free will could be shown not to exist, that wouldn't stop us from inventing it!

Many of us, I suspect, understand the foundations of the anti-free will argument well enough. At the very least we understand our will to be a tremendously constrained beast, easily knocked off its perch by alcohol, depression, or performance anxiety. We understand that in some conventional sense there is probably something illusory about the choices we make, and that at the heart of it all lie some ugly contradictions. But we don't sweat it at all. For free will is a tremendously useful illusion, and by that measure, as close to 'truth' as anything we will experience.

This raises an irksome question for the more zealous amongst the scientific community, who are wont to dismiss as irrational those who dare believe in anything other than the scientifically verifiable. Why bother getting het up about the choices made by automatons? By holding others accountable for the beliefs they choose to cling to, the zealots are themselves taking the greatest leap of faith of all. They are choosing to believe, despite a dearth of sensible evidence, in the existence of free will. Or as Karl Popper was fond of saying, when people try to convince me consciousness is an illusion, what exactly is it they think they are trying to convince?

We choose our beliefs for the most pragmatic of reasons, and sometimes this pragmatism demands we put consistency aside. Or in the words of the always eloquent Walt Whitman:

> Do I contradict myself? Very well, then I contradict myself, I am large, I contain multitudes.

8 The Little God of Pragmatism: Conclusion

Say what you like about fundamentalism, it sells. Nothing is quite as galvanising as a dose of righteous certainty. And that's a great shame.

David Hume had a delightful phrase, he said our thinking needed to retain 'a certain looseness, to stop the possible shrinking to the size of the actual.'

Immanuel Kant reminded us that we can never know the world as it is, but only the world as our mind is able to perceive it. Reality, reason, causation; these concepts are the products of our mind. We trust them because it suits us to trust them, because they have served us so well in the past. Their value then is utilitarian rather than transcendental.

This distinction between usefulness and truthfulness is as important as it is subtle. Unlike the Platonist I find no beauty in truth. Indeed I consider it an ugly word, a bludgeon wielded in battle by those who lack the social grace to back away. Yet all around us confident certainty abounds. 'Only through God can we discover the truth,' says the bully preacher. 'There is no God,' returns the angry biologist, 'Science alone can lead us to truth.' Well I say a plague on both your mystical houses.

My problem with fundamentalism, be it religious or scientific, is largely emotional. When it isn't boring me, it offends. I love a strong point of view, I enjoy being provoked and challenged in my thinking, but only by those who are happy to be provoked and challenged in return. People who can agree to differ, who are open to new perspectives, I find delightful. People who see a difference of opinion as an opportunity to save souls make my skin crawl.

To place a boundary around science is not to dismiss the glorious fun to be had within that perimeter. When Descartes suggested thought may be the only reality, for it is the only thing whose existence we can't doubt, he rubbed up against our common sense. The chair I sit on feels every bit as real to me as any thought I might have about sitting in it. When we read from Socrates 'the only thing we can know is that we do not know, and it is important that we know this', we may delight in the playful paradox, but instinctively we reject the conclusion. It feels to me as if I know many things. I know that rain is wet, that heavy things falling on my toe hurt more than light things, that marshmallow is softer than rock, and sweeter too, and that the pressure differential caused by a well-designed airplane wing can suck the craft up into the sky. None of this knowledge relies heavily upon cultural construction. Planes do not stay up in the air because of the collective belief systems of their passengers. No matter how much we scrutinise the motivations, biases and abuses of modern science, we mustn't throw the baby out with the bathwater. It may all be, in the words of Xcophanes, a woven web of guesses, but when it comes to the measurable and the predictable, some guesses are clearly better than others.

Ways of thinking which have exploited the existence of verifiable, objective measurement lie at the heart of the miracles of modern living. We can travel to the other side of the globe in a day, send photographs the same distance almost instantaneously by way of a humble cellphone, have cancerous tumours identified and treated, organs transplanted, or revel in the mundane but profound freedom of being able to jump in a car to visit friends and relatives (and later by the same means of locomotion, escape them). For most of human history, such things were simply impossible.

These achievements are the harvest of the scientific habit. The creative 'what if' spark of the hypothesis; the cautious, painstaking deduction that leads from there to a controlled experiment; the careful, critical art of measurement; the openness to falsification and the unshakeable belief in the physical world's regularity: these are the qualities that have produced predictions of remarkable accuracy, and technologies of commensurate power. In the competition for the modelling of the natural world, a clear winner has emerged, and only the fool maintains otherwise.

Thus it is plainly loopy to believe the world was made six thousand years ago by an old fella with a beard. When you see a medium, claiming to speak to the dead, of course you should ask yourself why they most often make contact with older men whose name begins with a … wait for it … I'm getting a strong J here … Jack … John maybe … and heart, I'm getting something about the heart, heart problems … no? Broken-hearted then… He left somebody close behind… Cancer, yes, I'm getting cancer…

This stuff is entertaining, perhaps, but to take it seriously is to wilfully reject the evidence before our eyes. We're being conned. We know this because such mediums are unable to outperform the predictive ability of self-confessed shysters who go up against them. In the same way, it is not true that Libras earn more money on average than Tauruses (sadly), homeopathy does not perform well in double blind tests, and people who are prayed for are no more likely to recover than those who are not. Along with the comforts of modern technology, science also delivers these at times discomforting conclusions.

In the same vein, those who tells us some aspects of the physical world (time, the emergence of life, consciousness) are forever beyond the reach of science because at heart their properties are too mystical, should be treated with scepticism. Maybe science never will get to the bottom of them, but the belief that it can is what has driven the discoveries of the past, and that's worth remembering.

Yet, as Huxley eventually discovered, science has its limits too. Not all important questions are scientific; not all the pressing issues facing us can be resolved by way of a wager. And this is where the zealot stands apart from the rational practitioner, for the zealot has no good sense of this distinction.

There are two clear areas in which science is of only limited use to us. The first is obvious and needs little explanation. Many of the situations we are called upon to deal with are simply too complex, and require too rapid a response, for the methodology to be applied. We are forced in our interactions with other human beings to trust our instincts, our folklore and the anecdotal experience we accumulate, because it's all we have available. What Christmas gift to choose for your partner, how honest to be at that job interview, how far to let the flirtation run, how much to save for your retirement? The world we live in is a messy, dynamic place, and our knowledge of it is at best probabilistic. This though is not controversial. Those who consult star signs or actuarial tables to negotiate their way through a marriage proposal we dismiss as lunatics or cowards.

Much more interesting, and easier to miss, are the questions which science struggles with not because of the limitations of our real time processing power, but because of the very nature of the questions themselves.

As a human being, I wish to do more than understand the world in physical terms. For the world I inhabit is patently not just a physical place. I may ultimately be made of fundamental forces, but I swim in a world of ideas. Think of it in terms of a television screen. Get close enough and you can see (with old style sets at least) that the picture is made up of thousands of little coloured dots. The drama you have been watching has in fact been an illusion, created by your mind in response to these cleverly pixelated suggestions. But while watching the little coloured dots may in some sense give you a truer picture of what is going on at the technological level, it is far less interesting and satisfying than viewing the illusion the dots create at a distance. 'What's on the TV?' someone asks. 'Dots,' you reply, smugly believing you have cut to the true heart of the matter. But you have overlooked the murder mystery which is keeping the rest of the room spellbound. 'No, you're all wrong,' you cry. 'There is no murder, there is no mystery. It really is all just dots!' And that makes you an idiot.

We can see ourselves simultaneously as forces, chemical reactions, living entities, cultural constructs, ideas. None of these aspects of our existence is more real or truer than any other. At some point we have to let the science go and simply tell ourselves stories, for it is from stories alone that we derive meaning, and without meaning we cease to exist as individuals; we collapse into dots. When I wish to make a judgement regarding beauty and value, I must turn to my personal stories. When I wish to make a moral choice in the world, science is of no use to me. Again, the stories I tell myself will guide me.

There is nothing flakey in this. The physical world is undoubtedly best modelled by the methodology of science, but the physical world needs to be interpreted before it makes any sense to us. In day to day terms, considering the world as a three-dimensional experience, filled with solid things and subject to the steadily ticking clock of time, is admirably sane. It may well turn out the three-dimensional model is deficient for some technical purposes, but to insist upon the primacy of this view is not to be commendably hard-headed, so much as pathologically abstract. I choose to believe that the idealists were wrong, and the world I perceive does actually exist, even if my perception of it may be faulty. I choose to believe that I have free will. I choose to believe others experience the conscious state in much the same way I do, that I am not a freakish exception in a world of zombies. And I choose to believe that some human actions, genocide for instance, are just plain wrong. These are all stories I embrace, despite them being for now unfalsifiable. This does not make me naive. It makes me human.

Even the most zealous rationalists choose to believe these things too, so at some level at least they understand the distinction. But like all mystics, they are mighty selective in their application of the rules. Thus they fall into the trap of claiming for science not just predictive, but also interpretative powers. And we encourage them, treating scientists as our new high priests; the role to which Huxley, who coined the term scientist, aspired. Nothing demonstrates more clearly how uncomfortable we have become with our own imaginations.

Setting up straw men is easy of course, so let me solidify this claim with an example, coming from perhaps the most famous of the modern day scientific evangelisers. Richard Dawkins wrote in his end notes to the 1989 edition of *The Selfish Gene*:

> … when you are actually challenged to think of pre-Darwinian answers to the questions 'What is man?' 'Is there meaning to life?' 'What are we for?', can you, as a matter of fact, think of any that are now worthless except for their (considerable) historic interest?

Well actually I can think of an answer. Or rather loads of them. Here's an example I quite like, courtesy of a chap you may have heard of, who had a bit to say about the human stain:

> Tomorrow, and tomorrow, and tomorrow
> Creeps in this petty pace from day to day,
> to the last syllable of recorded time:
> and all our yesterdays have lighted fools
> the way to dusty death. Out, out brief candle!
> Life's but a walking shadow; a poor player,
> that struts and frets his hour upon the stage,
> and then is heard no more: it is a tale
> told by an idiot, full of sound and fury,
> signifying nothing.

So Shakespeare has Macbeth lament, upon news of his wife's death. The mood is bleak of course, but the woods are closing in, the 'untimely ripped' executioner is on his way and poor old Macbeth is beginning to wonder if listening to those witches wasn't a tad foolish. So he turns his mind to exactly the questions Dawkins asked us to consider. What is man? Is there meaning to life? What are we for?

Macbeth's response is admittedly less than cheering, yet we can see he manages to capture the essence of the modern solution. Life is an elaborate pantomime, an acting out of our own invention, signifying only that which we ourselves imagine into existence. Knowing the world is beyond us, and the brevity of this existence our greatest challenge.

Yet Shakespeare wrote this in the time BD (Before Darwin) and so, by the philistine's view, he must be missing the magic ingredient of understanding that evolution brought. Within this context though, it's tremendously hard to see what that might be. It's not only Shakespeare who Dawkins would have us dismiss, but Aristotle too, and Confucius, Buddha, Jesus, Locke, Hume, Kant. The unspoken assumption is that only the coloured dots of the evolutionary process can help us make sense of the drama, which is a tragically anaemic take on our existence.

Darwin's *Origin* added something wonderful to the human narrative; it provided the first hint of the way we, in all our shambling, contradictory glory, assemble ourselves from star dust. Yes, there's poetry to be found there, if that's your bent. But to model physical relationships in this way is to tell only part of the story. To understand what we are, to accumulate the sort of knowledge required to pilfer meaning from our world, our thinking must be broader.

The philosopher Roger Scruton put it rather nicely:

> The scientific attempt to explore the 'depth' of human things is accompanied by a singular danger. For it threatens to destroy our response to the surface… It is in this thin top-soil that the seeds of happiness are sown.

Quite.

I have not set out to convince you that my way of approaching science is the right one, for 'convincing' suggests there is a single rational path from evidence through to conclusion, when of course no such thing exists. Our take on reality must in part be an aesthetic one. I've aimed to show you what the view from here is like, because I like it quite a lot. I like being able to enjoy the fruits of modern science without having to bow to any particular brand of interpretation. I enjoy being able to bring to bear on new discoveries the same sorts of scepticism I naturally apply to astrologers and politicians. I enjoy putting creativity at the very heart of my existence, believing that those things which most urgently require my human attention must be dealt with not by models, but by imagination. I enjoy the excitement of new discoveries as science expands its field of competence, and I like believing that the limit of such knowledge is in principle unknowable, that such deep mysteries as human consciousness may one day collapse beneath the objective gaze.

Most of all though, I love uncertainty. I love the freedom it gives me to engage with my world, and the value it brings to our storytelling tradition. Like all freedoms, the utilitarian stance comes with its attendant responsibilities. If your stories are different from my stories, and stories are how we reach our moral conclusions, then when it turns out we disagree on issues regarding crime and punishment, education, tax, reproductive technology or the environment, shouting 'I'm right and you're wrong' just won't cut it. We're going to have to listen to one another, and see if some sort of compromise can't be hewn from the rock of culture. The solutions we negotiate are going to depend upon our ability to find commonalities, the universal instincts that bind us, and they're going to make virtues of pragmatism and stability. We're going to have to respect the idea that the sacred is personal, and we're going to have to find ways of letting each other get on with our lives.

I think that's an excellent place to live.

■ ■ ■

Little Sophie is sitting on the front step in the mid-afternoon sun, and she is crying.

'Look,' says the evolutionist, 'the small child is engaged in an automatic response honed over millions of years. Her shrieking has co-evolved with the mother's sensitivity to the child's cry, both a warning mechanism and the means by which offspring actively compete for their share of the scarce child-rearing resources. That is why Sophie is crying.'

'Not so,' says the biochemist. 'For you evolutionists make the mistake of looking only at the surface for your explanation, and so your tale lacks the richness of the truly wonderful human machine. Sophie's crying is in fact the result of a complex and beautiful series of chemical reactions, beginning with the electrical impulses sent to the brain as a result of sensory inputs, and then asserting themselves as a delicate rebalancing of the chemical composition inside her head. Signals then travelling back from the brain are setting up changes in the chemical conditions of the individual cells and the crying you are witnessing is in fact just glorious chemistry, the making and breaking of molecular bonds.'

'Oh, you chemists and you're damned oversimplifying,' harrumphs the physicist. 'Why must you insist upon painting every picture with the colours of your clumsy atomic models? For what use is it understanding that chemicals are reacting with one another if we do not understand why? The reactions you see in Sophie are the result of the very nature of the forces which define the existence of the fundamental particles themselves. For it is these forces which give the chemical reactions you are so proud of their special characteristics.'

'Phooey,' says the radical String Theorist. 'I tire of these three-dimensional models of the world. To say that Sophie is crying is to assume that an event is occurring, but for an event to occur we must assume the existence of time. Time however is an illusion which pops out of the elegant mathematical equations which define the two dimensional world we actually inhabit. If you want to understand why Sophie is crying, you need to first define the equations which create the illusion of Sophie's existence in three-dimensional space.'

'Look, Mummy,' interrupts Leroy, Sophie's brother. 'Sophie's ice-cream has fallen onto the ground, and now she's crying.'

Sophie is crying because a spell has been broken; a moment of sugary, creamy togetherness has been split asunder. Leroy knows that the only way to stop Sophie crying is to give her the rest of his ice-cream.

So young Leroy, only four, faces a formidable moral choice. It wasn't his ice-cream that fell. He wasn't that careless. Then again, she is his little sister, and she is crying, and he can make it all right. He thinks about this, takes one last delicious lick and passes over the dripping cone. The crying subsides. The sister looks up and smiles.

Such is the art of living.

9 On the Shoulders of Giants: Bibliography

Below is a list of the books from which I scavenged my information, inspiration and confusion. (Yes, I know bibliographies are supposed to be alphabetical. Think of this as a random walk through the library.) I've listed them according to the chapter they most contributed to, and have tried to give a very brief taste of what you will find within their pages. I hope some of you will follow the trail.

Chapter One

Three books in particular got me thinking about the big question in Chapter One. What is scientific thinking, and where do its limits lie?

***Consilience: The Unity of Knowledge*, Edward O. Wilson**
(New York: Abacus, 1998)
Edward Wilson kicked off one of the big modern controversies in the seventies, with his idea that the study of biological systems could shed light on the human condition. In this book, he goes the whole way, essentially arguing that the scientific way of thinking is the best tool for understanding all aspects of human existence. That's a big call, and his argument ultimately falls short of its own publicity, but it's a fascinating, provocative read: evidence that sometimes the books we disagree with are the most valuable reading experiences.

***The Hedgehog, the Fox and the Magister's Pox*, Stephen Jay Gould**
(London: Jonathon Cape, 2003)
This book was written as a rebuttal to the *Consilience* argument. Gould maintains that scientific endeavour has limits, and beyond its edges we

must turn to the artistic and philosophical for our guidance. This is an idiosyncratic read, the line of reasoning buried deep beneath a tangle of discursion. Frustrating, in that way, but charming too. Doesn't quite deliver *Consilience* the killer blow it intended, perhaps because in the end this isn't an either/or argument.

Evolution as a Religion: Strange Hopes and Strange Fears,
Mary Midgley, rev. edn (London: Routledge, 2002)
Midgley brings a philosopher's perspective to the debate, using evolutionary theory as her case study as she tries to show how easy it is for scientists to be lured into metaphysical speculation. There are some good examples of scientific abuse in this book, and it is nice to see the topic approached from the perspective of big ideas. Midgley touches upon, but does not fully articulate, an alternative path to answering the big questions, and as with Gould, this lack of detail leaves one wondering if the treatment isn't a little one-sided.

Chapter Two

I have no real background in philosophy, and so had to rely upon the erudition and clarity of some fine writers to get my head around the sorts of issues I wanted to address in this chapter. Some of the best introductions are below:

The Shorter Routledge Encyclopedia of Philosophy, **ed. by Edward Craig**
(London: Routledge, 2004)
This weighty tome served as a wonderful reference, a great place to check that a particular piece I'd come across didn't represent too unorthodox a point of view, and to hunt down definition and background detail. Because this edition has worked at cutting down on the number of references, so as to leave the major contributions unabridged, it's also an excellent place to go for well written and considered introductions to the philosophical greats.

The Great Philosophers, **ed. by Ray Monk and Frederic Raphael**
(London: Phoenix, 2000)
This is a collection of essays on some of the big names in Western Philosophy. They are not overview essays, but rather in each one, an acknowledged expert is given a chance to highlight a particular aspect of a great philosopher's work. The essayists mostly quote heavily from the original works, giving a

flavour of the style of the thinker in question. I drew heavily on the pieces on Plato, Descartes, Hume and Popper in this book, and only wish they'd slipped Kant into their line up.

A History of Western Thought, **Gunner Skirbekk and Niles Gilje**, 7th edn (London: Routledge, 2000)
This one is more of an introductory text, giving brief, clear introductions to the important thinkers within this tradition, and trying to put their contributions into the broader context of what philosophy is all about. An easy read, and great place to go for the basics, or for a simplified summary when the source is making your head hurt.

History of Western Philosophy, **Bertrand Russell**
(London: Allen and Unwin, 1946)
How it is and what it all means according to that charming, opinionated, irascible rogue, Bertrand Russell. An important philosopher in his own right, here Russell gives his own interpretation of all that had gone before him. Self-serving of course, but that's part of the charm, and the strong opinions are somehow so much easier to digest when the subject matter is complex. Russell is a fine communicator, and has the logician's knack for cutting to the heart of issues. There's plenty here to make you chortle too, something of an achievement in the field.

Common Sense, Science and Scepticism, **Alan Musgrave**
(Cambridge: Cambridge University Press, 1993)
This is the book I found the most helpful when it came to getting my head around Hume, Kant and Popper, and the relevance of the induction challenge to science. Because it is philosophy filtered by the needs of a particular theme (the importance of scepticism and its challenges) there is a nice logic to the way this story unfolds. And Musgrave is a fine teacher, distilling the key points and helping you understand their importance.

The Enlightenment, **Norman Hampson**
(Harmondsworth; New York: Penguin, 1968)
Just one of those books I stumbled upon. A summary of eighteenth-century thought in Europe, and valuable in that it is not considering the issues from a scientific perspective, so there is a broader context to draw upon.

A Short History of Modern Philosophy, **Roger Scruton**, 2nd edn (London: Routledge, 1995)
Another succinct and powerful summary of some of the key thinkers from the Enlightenment period on. Notable for its championing of Kant as the key figure in modern philosophy, it's a nice counterpoint to the dominant thinking amongst most philosophers of science.

Modern Philosophy: An Introduction and Survey, **Roger Scruton** (London: Pimlico, 2004)
Maybe the best of the philosophy books I have come across, and a wonderful companion piece to Russell in that it picks up on the debates of the last century and leaves one feeling at least a little more up to date. This book works through themes of enquiry rather than the contributions of individual philosophers, which means you are particularly reliant upon the author's own interpretation. In the end this feels much more a strength than a weakness, thanks to the clarity and eloquence of the writing.

Science: A History 1543–2001, **John Gribbin** (London: Penguin Books 2002)
This is a delightful read, well-informed and excitable, as good history should be. Here is the place to go to fill in the details of the quite remarkable adventure that is the path of modern science. You'll be stunned by the cleverness of those who first had the ideas from which our modern lifestyle is assembled, and better still you'll be entertained by the many diversions into their personal stories. Interestingly, Gribbin is no fan of Kuhn's paradigm shifts, and prefers instead to highlight the incremental, stepwise path to modern knowledge. With more examples and details on offer than in Kuhn's work, one is left wondering whether Kuhn may have indeed overstated his case.

Chapter Three

The best way to find out what Popper was on about was to read him, I figured. The stuff I found most interesting is listed below, along with a wonderful contribution from Michael Ruse.

The Open Society and its Enemies, **Karl Popper**
(London: Routledge, 1945)
Although not at this stage concerned with the question of science and the scientific method, here Popper sets out his concerns regarding the dangers of historicism, and in particular what he sees as Plato's unfortunate legacy. Many of Popper's key concerns come through here: the uncertainty of knowledge, and the need therefore for tolerance, dialogue, scepticism and stepwise pragmatic solutions. Perhaps due to my own prejudices on these matters, this feels like an important book, as relevant today as it was in the postwar environment.

Conjectures and Refutations: The Growth of Scientific Knowledge,
Karl Popper (London: Routledge and Kegan Paul, 1963)
Although *The Logic of Scientific Discovery* is the place to go to find Popper's original discussion of the scientific method, this is the book I found most helpful with regard to that topic. A series of essays and lectures, it is written with a broader perspective and greater need for clarity than the other works. Here Popper clearly lays out his theory regarding the way science progresses, and at the same time considers the nature of philosophy itself, and many of the issues which surrounded modern philosophy at the time. The fact that this came later means that here he also addresses many of the issues which arose as a result of the reaction to *The Logic...* A hugely confident writer, Popper is always worth reading, albeit with your own scepticism on high alert.

Knowledge and the Mind-Body Problem: In Defence of Interaction,
ed. by M.A. Notturno (London; New York: Routledge, 1994)
Again not strictly on the science thesis, but an interesting look at the way Popper approached the issue of the mind and dualism, as considered in chapter seven, and a further demonstration of his method of 'common sense' philosophy. A nice antidote for those who fear the materialist agenda, this series of lectures from the early seventies introduces Popper's Three World's theory, a forerunner perhaps of Dawkins' later flirtation with cultural evolution. I don't myself buy Popper's dismissal of materialism, but it is interesting to wonder what he would have made of late century advances in neuroscience, computing and genetics.

The Structure of Scientific Revolutions, **Thomas Kuhn**
(Chicago: University of Chicago Press, 1962)
Having read so many interpretations of Kuhn's work, it came as something of a surprise to see how often he has been misrepresented, by supporters and critics alike. This is an interesting and thoughtful stroll through some of the history of scientific discovery, all packaged together within the structure that has since become famous, but there isn't quite the explanation of progress you might expect.

Mystery of Mysteries: Is Evolution a Social Construction?
Michael Ruse (Cambridge, MA: Harvard University Press, 1999)
This book is an excellent way of examining the Kuhn/Popper debate, using the history of evolutionary thought as its case study. You get the philosophy, the history and the science all for one price. Ruse, a philosopher with a particular interest in evolutionary theory, keeps it clear and interesting throughout.

Chapter Four

Here the book list really starts to pile up. There are so many people writing on the topic of evolution in a clear and arresting manner that it's hard to know where to start. So heavily populated is the field, I won't pretend these titles offer the best insights: they're just the best of the ones I found time for.

The Blind Watchmaker, **Richard Dawkins** (London: Penguin, 1986)
Whatever you might think of his often forthright views, Dawkins remains the master communicator in the field of science. More lay people have managed to get their head around evolutionary theory thanks to him than through any other source. And *The Blind Watchmaker* is, for my money, Dawkins at his best. His clarity is not achieved at the expense of detail. Like all the very best elucidators, his great gift is to make the reader feel smart. If you're wondering how natural selection does its business, start here.

Darwin's Dangerous Idea: Evolution and the Meaning of Life,
Daniel Dennett (New York: Simon and Schuster, 1995)
To call Dennett an enthusiast for natural selection would be to understate things. A philosopher with a passion for communicating his ideas to a wider audience (brave and rare) he takes us through how natural selection works,

and what, to his mind, it means for all of us. Unashamedly putting storytelling ahead of science, he goes for the big picture, interpretation and implication. He's not afraid of launching personal attacks, most notably in a weird chapter on Stephen Jay Gould, and that's the book's greatest weakness. You don't have to agree with him though, and he does force you to engage. It's a great strength, and makes this an exciting read.

The Ancestor's Tale: A Pilgrimage to the Dawn of Evolution,
Richard Dawkins (Boston: Houghton Mifflin, 2004)
I could easily fill this list with Dawkins titles but decided on just two. This more recent work paints the tale of life on a broad canvas. A fascinating piece of writing, taking us from modern day humans all the way back along our particular branch of life's tree, stopping at every intersection to greet relatives and forebears along the way. As you'd expect from Dawkins, it's crammed full of delightful tales from the natural world. His enthusiasm and boosterism is not to everybody's taste, but he is toned down on this occasion. Treat him as a favourite uncle whose excesses are tolerated for the entertainment and energy he brings to the party, and this a fine, informing read.

Origins Reconsidered: In Search of What Makes Us Human,
Richard Leakey and Roger Lewin (New York: Doubleday, 1992)
The tale of human evolution, as told by those getting their hands dirty on the ground. This bone digger's perspective gives a valuable insight into the way the clues are put together, the storytelling involved and the way feuds can railroad science.

Huxley: The Devil's Disciple, **Adrian Desmond**
(London: Michael Joseph, 1994)
From its cover you might expect a dense, scholarly biography. It's certainly thick enough, and meticulously researched. That said, there is enough life in Desmond's prose and Huxley's tale to make this a pleasure, not a chore, to those of us who find the details a little daunting. My brief summary of Huxley's life may read like a convenient exaggeration, but read this and you'll see where my interpretation comes from. This book gives you a great feel for the details of the time, and the impact evolution had on Victorian society.

Dawkins vs Gould: Survival of the Fittest, **Kim Sterelny**
(Cambridge, England: Icon Books, 2001)
Evolution is full of scraps, and the animosity that existed between Gould and Dawkins is well known. This study, by philosopher Kim Sterelny, gives a useful outsider's perspective, and shows nicely how two big brains can spin such different stories from the same set of facts. Although the argument is in essence peripheral to evolution's story, it is fascinating to see the way viewpoint influences interpretation.

Science on Trial: The Case for Evolution, **Douglas Futuyama**
(Sunderland, Massachussetts: Sinauer Associates, 1982)
Still considered by many the classic popular refutation of the creationist case, the great strength of this book is that it argues by carefully elucidating the case for evolution, and providing all the little dates and details needed to answer the usual protestations. Even if the creationism debate isn't of interest, this still stands as an elegant introduction to evolutionary theory. The only weakness, I think, is that it perhaps falls for the well-laid trap of concentrating on the detail of the creationist argument, so neglecting to fully address the weaknesses in the logical construction of the case.

The Trial of Evolution, **Niles Eldredge**
(New York: Nevraumont Publishing, 2000)
Another look at the evolution/creation debate from an evolutionist's perspective. Because Eldredge is famous for his work with Gould in questioning some of the assumptions made by natural selection enthusiasts, he is upon occasion name-checked as one who has cast doubt upon the evolutionary edifice. Angry at this misrepresentation, he here sets out to put the record straight.

Darwin's Black Box, **Michael Behe** (New York: Touchstone, 1998)
You should read this book, if only to see what it is the so called Intelligent Design debate is about. There are of course loads of books expounding the virtues of creationism, but I've never been able to stick with any of them. They rely too heavily upon the art of the bare-faced lie, and that's a little disheartening. Behe's attack is more genuine I think, and deserves to be considered. As I said in this chapter, at its heart lies a logical rather than a technical flaw, which is worth remembering, for this book is mostly about the technical stuff, allowing the logic of the case to go unexamined.

The Evolution-Creation Struggle, **Michael Ruse**
(Cambridge, Massachussetts: Harvard University Press, 2005)
Following on from the Huxley debate, this book takes a broad, philosophical view, and asks interesting questions about the value of positions behind both sides of this debate. It's refreshing to see someone prepared to take the wider perspective on this. If you've ever wondered why some churches find evolution threatening, and others accept it with a shrug, why some scientists are enraged by creationists, and others are more sanguine, this is a great place to start. It's the story of the ideas behind the idea, I suppose. Remarkably helpful when it comes to thinking about the interactions between science and story.

Chapter Five

Everybody wants to read about modern genetics and the link to human nature, it seems. What makes us the way we are? It's an enticing question. Scientists are relishing the chance to contribute to a debate in which the general public are genuinely interested for a change. At times this enthusiasm to engage has perhaps compromised their methodology, but critically attuned readers will find loads that stimulates in these books. In terms of the quality of prose, some of the very best explanatory writing you would ever hope to come across can be found here.

The Selfish Gene, **Richard Dawkins**, 2nd edn
(Oxford: Oxford University Press, 1989)
Here is the book that more than any other got the whole debate into the public arena. How does the reproductive ambition of the gene impact upon the development of the life forms we are familiar with? The answer to this question introduced a new perspective, and a memorable phrase, to the debate. The book is also a superb way of getting your head around the natural selection mechanism. As noted before, Dawkins is an enthusiast, so 'hear what he says and divide by two' is a pretty good rule. The chapter on memetics shows that you can push a good idea a little too far, and the reader needs to constantly distinguish between model and metaphor. Still that is a responsibility we must accept, if we wish to seriously engage with such arguments.

DNA: The Secret of Life, **James D. Watson**
(London: Random House, 2004)
The story of modern evolution as told by one of its key players, the co-discoverer of the double helix. Watson is not one to pull his punches and is rarely moderate in his opinions; but that said there is far more history and impressively readable technical explanation than polemic here. One of the best overviews of the field of genetics I have come across, with loads of historical intrigue to keep it spicy.

The Red Queen: Sex and the Evolution of Human Nature, **Matt Ridley**
(London: Viking, 1993)
Matt Ridley is, for my money, the best plain-English explainer of evolutionary theory anywhere. As a non-practitioner he has licence to be more whimsical than Dawkins, and plays shamelessly to the crowds. This book is a lively introduction to sexual selection theory, filled with interesting snippets with which to bore your friends. The two things to be wary of are that Ridley often writes with airy certainty, where other sources suggest more caution, and the leap from animal to human applications is made with more confidence than the material can support. But again, read with caution and it is truly wonderful, informative writing.

Genome: The Autobiography of a Species in 23 Chapters,
Matt Ridley (London: Fourth Estate, 1999)
What is true for *The Red Queen* is true here. Superbly written, great explanations, and an absolute wealth of amusing anecdote and detail. This book takes the human genome one chromosome at a time, highlighting the discoveries piling up in terms of what bits of development the various genes contribute to. Gives a great sense of the medical potential of this sort of work, although again, enthusiasm perhaps gives way to misguided confidence at times.

Nature via Nurture: Genes, Experience, and What Makes Us Human,
Matt Ridley (London: Fourth Estate, 2003)
A curious book this one. Tremendously successful, it leaps into the speculative, controversial world of human development. Nature or Nurture? The verdict's in, Ridley tells us in his excited way. It's both! Well yes, but we already knew that didn't we? I think, if you trace this argument carefully, there's something sneaky going on. Setting himself up as the model of bal-

ance and compromise, Ridley than goes on to push the nature side of the argument fairly hard. This is best seen with his treatment of twin studies. Absolutely nothing wrong with having an opinion, but writers like Dawkins and Dennett, who wear their intentions on their sleeve, get more brownie points from me. With Ridley you have to dig a little deeper to find the perspective being defended. That said, it's as good as it gets on the explanatory front, and a sensible reader can work these things out for themselves.

***The Language Instinct*, Steven Pinker** (London: Penguin, 1994)
This is my favourite of Pinker's books. Here he takes an area he knows very well — language — and uses it to suggest the way we learn is as dependent upon inherited thinking structures (what we might think of as our innate language skills) as it is on the things we pick up from the environment. If this hypothesis is right, and it is strongly argued here, then it gives us a great metaphor for intellectual development in general, a theme Pinker explores in his later books.

***The Blank Slate: The Modern Denial of Human Nature*, Steven Pinker** (London: Penguin, 2002)
This book is Pinker at his most strident, and that stridency I think is a shame. Many of those who wish to attack the evolutionary hypothesis when it comes to human nature choose their examples from this book, and you can see why. As always with Pinker, this is well researched and forcefully written, but at heart the whole thesis depends upon the opposition really saying the things he claims they are saying (specifically that human nature is entirely dependent upon its environment). I haven't been able to find these claims in the work of his critics. Without them the book falls a little flat.

***Dear Mr Darwin: Letters on the Evolution of Life and Human Nature*, Gabriel A. Dover** (Berkeley: University of California Press, 2000)
This book isn't so well written, to be fair. In fact, its central conceit, an imaginary correspondence between Darwin and the molecular biologist Dover, is a little embarrassing. Stylistically, Dover is over-reaching. What's more, he's angry, very angry indeed. This book wants to show that Dawkins and his ilk are on the wrong track entirely, and he's not above calling them names to make his point. In a way that's a shame, because beneath these glaring flaws there's a lot of interesting and important stuff. He at the very least

establishes a good technical case to answer. It's a shame his communication skills don't quite match those of his opponents, for it would be sad to see the debate resolved on those terms alone.

***Alas Poor Darwin: Arguments Against Evolutionary Psychology*, ed. by Hilary Rose and Steven Rose** (London: Vintage, 2001)
This is more like it. Dover himself is a contributor to this collection of essays, and within the more constrained format manages to make more sense. The aim of this collection is to put the evolutionary psychology agenda under the spotlight, and with a range of writers with a range of agendas the result is a little patchy. Theirs is the moral high ground, and I wonder if in a funny way this constrains the writing. Having accused the other side of wild exaggeration, the writers here are often forced to adopt a restrained tone, and at its worst this leads to a sort of pomposity. The devil gets all the best lines, as they say. That said, there is an important case to answer, and the best of these essays call it to account. It's a must-read for anyone who has enjoyed the shots fired from the other side.

***Lifelines: Beyond the Gene*, Steven Rose,** rev. edn (London: Vintage, 2005)
Frustrated at the success of the reductionist agenda, and the great popularity of the selfish gene perspective, Rose here attempts to introduce an alternative metaphor into the discussion. In *Lifelines* the author tries to show how a human life is best understood not simply as the outcome of an evolutionary struggle, but also as each individual's unique interaction with its personal environment, culture and history. The model being offered is undoubtedly more subtle and complex than one in a work by Pinker for example, but sadly it is exactly this complexity which denies him the common touch. Rose is too careful a thinker to shelter beneath slogans, the downside being that the reader has to work hard to distil what is a tremendously important message. If this book were as widely read as some of the others mentioned in this chapter, the public debate would be a far more interesting one.

***Sense and Nonsense*, Kevin Nalad and Gillian Brown**
(Oxford: Oxford University Press, 2002)
This book is a good overview of the field of evolutionary psychology. It surveys the various approaches and tries to give an account of their strengths and weaknesses. It is skewed of course by the fact that the authors are practition-

ers in the field, but it's a far quieter, more reasoned consideration of science's contribution to our understanding of human nature than most of the populist tracts. Reading this feels like taking a moment out of the heat of the day in a cool room, with a gentle breeze and the quiet voice of reason in the background. Not that I bought its main conclusions. See what you think.

Chapter Six

There are fewer books referenced in this section, and in hindsight that's probably because I was amongst people working in this field during the year so it was easier accessing reports and first-hand opinions. Also, as it's a more specialised area of interest, there isn't as much out there aimed at the casually curious. And, to be honest, I smile more reading science and philosophy books than those on history. Each to their own. That said, here's a title or two which I did work my way through in order to learn a little more about Pacific migration.

***The Lapita Peoples: Ancestors of the Oceanic World,* Patrick V. Kirch**
(Cambridge, Massachussetts: Blackwell, 1997)
There's a whole heap here on what we know about the Lapita people and how we've found it out. Loads of interesting stuff on how cultures are defined and how the clues are pieced together to flesh those definitions out.

***The Prehistoric Exploration and Colonisation of the Pacific*,**
Geoffrey Irwin (Cambridge: Cambridge University Press, 1992)
The best thing about this book is its in-depth look at the potential sailing and navigational techniques used by the Pacific's early explorers. If you're interested in how on earth anyone without a compass or GPS could make their way about the vast expanse of the Pacific Ocean, managing to find all the little specks of land along the way, have a read of this. As with the Kirch book, the style isn't particularly racy, but the story's a good one, which compensates.

Where the Waves Fall: A New South Sea Islands History
***from First Settlement to Colonial Rule*, K.R. Howe**
(Sydney: George Allen and Unwin, 1984)
If you're looking for a single book to summarise the broad sweep of Pacific history, with an emphasis on post-European contact times, this will give you a decent sort of an overview.

The Quest for Origins: Who First Discovered and Settled New Zealand and the Pacific Islands? **K.R. Howe** (Auckland: Penguin, 2003)
A good read for anyone wanting an up to date discussion of how people found their way to this part of the world, and how we know. It's done in the context of what we used to think, and how our stories drive our models. At the end there's a very short discussion of the way modern mores might be skewing the current story. It would have been interesting to me to see this idea developed more, but maybe that's the next book.

Guns, Germs and Steel, **Jared Diamond** (London: Vintage, 1998)
Diamond specialises in the grand theory, the history of the world as explained by food groups in this case. I don't have the historical knowledge to know how this one measures up against the competing explanations, but it is clear from the way this case is made that competing explanations there will be. This is an ingenious bit of type-two construction. Can we use the global distribution of flora and fauna to explain the story of human progress? It turns out we can, if we squint through one eye and the wind is just right. Fun watching how the case is compiled, and like so many of these wonderful books, cram-packed with the sorts of facts and stories that you will be annoying others with for months. Included in this section because Diamond, having strong Pacific links, gives this area of the world much more consideration than most broad histories.

Chapter Seven

The mind is a tricky thing, but the very best writers can help us understand the basics without having to get ourselves another degree. This is true both in the scientific and philosophical fields. Some of these books are simply stunning.

The 21st-Century Brain: Expanding, Mending and Manipulating the Mind, **Steven Rose** (London: Jonathon Cape, 2005)
Yes, the same Steven Rose behind *Alas Poor Darwin*, and the earlier *Not in our Genes*, so we can make all sorts of assumptions about his agenda if we wish. This book though is less about that, and more about explaining how the brain does its job, at the level of neurons, synapses and chemical interactions. This is Rose's speciality. He is not just a commentator, but also a prac-

titioner in the field, and as is often the case, the more a person knows about the brain, the more cautious they become. He talks about how the brain might have evolved over time, covers ageing, mental illness, medication and possible future developments, and displays little taste for the philosophical quagmire of the mind. A restrained, informative piece.

How the Mind Works, **Steven Pinker** (London: Penguin, 1997)
I'm not sure many people have ever accused Pinker of restraint. He has a massive enthusiasm for his material, and a burning passion to get it across. He knows it, he's excited about it, he wants you to know it too. Like Rose, he steps neatly around the big philosophical issues and concentrates on how the brain does its thinking. Unlike Rose though, and this probably reflects his different academic background, he is doing his analysis at a slightly coarser level. There's less talk of chemistry, and more of why optical illusions work the way they do, or what happens when the walls shake in a room you assumed to be stable (you fall over, even though the floor doesn't shake). On this sort of amusing and illuminating example, Pinker is without peer, and it's hard to match his sheer energy; you are assaulted by the prose. You'll learn a lot about the tricks our brains rely on to get us through the day, reading this book. When he gets onto his evolutionary psychology hobby horse, the explanations become less rigorous, the gaps require bigger jumps and the scale of the title's ambition becomes a little more apparent.

Consciousness Explained, **Daniel Dennett** (London: Penguin, 1991)
Speaking of ambitious titles ... I've got a soft spot for Dennett's writing, although he drives some people mad. He is certainly combative. He's not looking for compromises, he's looking to define the fight. Traditionally many philosophers have held that the mind is unknowable by its very nature. Dennett isn't so sure. He has a hunch that modern science, and modern computing advances, will reveal that there's less to the mind metaphor than meets the eye. There's a wealth of information and speculation here, some of it scientific, some of it philosophical, and all of it presented in the style of a bar room brawler, doing his utmost to lure his opponents into throwing that first punch. I'm not yet convinced by his model of consciousness, but that's hardly the point. He's showing traditionalists they have a case to answer, and that I can only admire.

Growing up with Lucy: How to Build an Android in Twenty Easy Steps, **Steve Grand** (London: Weidenfeld and Nicolson, 2004)
I just had to include this title because it's such a different book. Grand isn't interested in theory for theory's sake. He's trying to build himself a robotic brain. Not a computer that gives a facsimile of thinking, but one that actually uses the processes we use, or that he supposes we use. Grand, in the tradition of windmill tilters everywhere, believes that traditional artificial intelligence programmes are on the wrong track. He explains why, and takes us through the trials and tribulations of robot building. Surprisingly technical in parts, this may be the work of a madman but I for one am not qualified to make that call. It's a cool little read.

The Private Life of the Brain, **Susan Greenfield** (London: Penguin, 2002)
Another neurologist, Greenfield sets out to explain the link, as she sees it, between rational, considered thought, and our emotional lives. The metaphor she uses is that of the network, and given her expertise, she is able to build a convincing picture of the way neural networks build up during our lives, so profoundly changing the way we subsequently experience the world. We are, in the end, defined by the connections our brains have formed, she argues. Great stuff here on humour, sadness, drug use, childhood and ageing, and important reminders from one who should know that the relationship between genes and environment is deeply, deeply complex and unique. Greenfield too wimps out on the big philosophical issues. Why does consciousness feel the way it does? Maybe we'll never know she says. It's interesting how workers in the brain sciences cut themselves off so cleanly from their philosophical backdrop.

Sweet Dreams: Philosophical Obstacles to a Science of Consciousness, **Daniel Dennett** (Cambridge, Massachussetts: MIT Press, 2005)
Dennett is back, and this time it's all philosophy. In a series of predictably provocative essays, he lays out again the case for ignoring those who argue that the consciousness problem is just too hard. Lots of talk of Colour Scientists, Chinese Rooms and zombies. I loved this book. It made my head hurt, and I'm a bit of a sucker for the frustrated, indignant style.

The Mystery of Consciousness, **John Searle** (London: Granta, 1997)
Searle is the anti-Dennett. Openly, angrily so. This is a feud good and proper, and so it's important to get a taste of the argument from both sides. Here Searle explains why it is that the mind must have an unknowable quality. A good companion piece to the title above, read it for yourself and see if you can negotiate a comfortable path through the tangle.

Mind: A Brief Introduction, **John Searle**
(Oxford; New York: Oxford University Press, 2004)
This is Searle's attempt to resolve the mind/body debate by defining it out of existence. I might be missing something of its subtleties, because it seems to me he is perilously close to coming over to the other side. Nevertheless, he continues to profess his deep attachment to The Chinese Room puzzle, so in his mind at least, the feud continues.

Freedom Evolves, **Daniel Dennett** (New York: Viking, 2003)
Maybe, in the end, this whole mind thing does come back to the stories we tell ourselves about our own free will, about what it means to be a conscious, motivated and responsible agent. Dennett thinks it does, and here attempts to show that we can be free agents in a deterministic world, and that the apparent contradiction is, as ever, a definitional one. His strident tone can become tiresome, and this isn't the clearest of his books, but again, he has a knack for putting science into its philosophical context.

Some Extras

Although I mostly relied upon evolution to provide me with examples, at times I delved into other areas of science, and in doing so I relied heavily upon the following titles.

Galileo's Finger: The Ten Great Ideas of Science, **Peter Atkins**
(Oxford; New York: Oxford University Press, 2003)
Atkins sets out to explore the state of modern science by considering some of the big ideas underpinning our current understanding. So if you want to know a little more about thermodynamics, relativity, string theory, quantum mechanics or modern mathematics here's a good place to turn, with

one proviso. This is tricky stuff, and Atkins is attempting to get the material across without oversimplifying. So your brain will feel well exercised, and if you're anything like me, bits of it will still seem mysterious on the third reading.

Schrödinger's Kittens and the Search for Reality, **John Gribbin**
(London: Phoenix, 1995)
All you ever wanted to know about quantum physics, and a little more besides. Gribbin is one of the big names in science writing, and the breadth of his experience allows him to provide the historical context in which the questions posed by the quantum world best make sense. Gribbin has his own interpretation on what is going on here, which the whole book builds up towards. Even if this part leaves you cold, it is still a superb, accessible introduction to a baffling field.

The Fabric of the Cosmos: Space, Time, and the Texture of Reality,
Brian Greene (London: Allen Lane, 2004)
The theory of relativity, quantum mechanics and string theory. These should be too complex to be adequately conveyed to a lay audience, yet somehow Greene manages to pull out the interesting stuff without resorting to dense equations. A genius communicator. If I had to choose a single work that convinced me that any idea, no matter how difficult, can still have its essence conveyed to anybody interested enough to listen, this would be it. An inspiring note on which to end.

Index